Factorization Methods for Discrete Sequential Estimation

This is Volume 128 in
MATHEMATICS IN SCIENCE AND ENGINEERING
A Series of Monographs and Textbooks
Edited by RICHARD BELLMAN, *University of Southern California*

The complete listing of books in this series is available from the Publisher upon request.

Factorization Methods for Discrete Sequential Estimation

Gerald J. Bierman
Jet Propulsion Laboratory
California Institute of Technology
Pasadena, California

ACADEMIC PRESS New York San Francisco London 1977
A Subsidiary of Harcourt Brace Jovanovich, Publishers

ACADEMIC PRESS, INC.
111 Fifth Avenue, New York, New York 10003

United Kingdom Edition published by
ACADEMIC PRESS, INC. (LONDON) LTD.
24/28 Oval Road, London NW1

Library of Congress Cataloging in Publication Data

Bierman, Gerald J.
Factorization methods for discrete sequential estimation.

(Mathematics in science and engineering series)
Bibliography: p.
Includes index.
1. Control theory. 2. Estimation theory.
3. Digital filters (Mathematics) 4. Matrices.
I. Title. II. Series: Mathematics in science and engineering.
QA402.3.B5 519.5′4 75-40606
ISBN 0-12-097350-2

PRINTED IN THE UNITED STATES OF AMERICA

To my colleagues
whose interest in square root estimation
motivated this monograph.

Contents

VI Inclusion of Mapping Effects and Process Noise

VII Treatment of Biases and Correlated Process Noise

VIII Covariance Analysis of Effects Due to Mismodeled Variables and Incorrect Filter a Priori Statistics

IX SRIF Error Analysis of Effects Due to Mismodeled Variables and Incorrect Filter a Priori Statistics

X Square Root Information Smoothing

Preface

Kalman's elegant solution of the discrete linear estimation problem has had a major impact in academic engineering circles and in the aerospace industry. Recently, areas of application have widened to include such diverse subjects as transportation planning and scheduling, marine and satellite navigation, power systems, process control, surveying, earthquake prediction, communications, economic forecasting and analysis, water resource planning, human modeling, and biomedical applications. Unfortunately, numeric accuracy and stability problems have often prevented researchers in these diverse disciplines from successfully computer mechanizing Kalman's algorithms. It has come to be realized that algorithm formulations that involve matrix factorization generally enhance numeric characteristics and at the same time generate new analytical perceptions. Our aim in this monograph is to familiarize analysts in these various areas with those matrix factorization techniques that lead to estimation algorithms that are efficient, economical, reliable, and flexible. In order to expedite their application we have included FORTRAN descriptions of the important algorithms. Our experience leads us to conclude that a better understanding of the underlying mathematics involved with matrix factorization, together with an awareness of the concise FORTRAN implementations that are possible, will result in a more widespread utilization of this technology. Hopefully, after familiarizing himself with this material, the reader will share our opinion.

This monograph came about as a result of a set of lectures on least-squares estimation that I gave at the Jet Propulsion Laboratory. The interest with which these lectures were received convinced me that the estimation applications community would benefit from, and in fact was in need of, an estimation reference book that thoroughly describes those matrix factorization methods and associated numerical techniques that have been successfully employed by numerical analysts.

Emphasis in this book has been on the algorithmic and computational aspects of discrete linear estimation. Although our material is of a rather specialized nature, it is of interest to a diverse audience. Our pragmatically oriented and detailed presentation will, we hope, make this book a useful reference.

Acknowledgments

It is difficult to acknowledge properly all those responsible for the work described here. Nevertheless, I would like to express appreciation to JPL management, particularly Dr. D. W. Curkendall and Dr. W. G. Melbourne, for encouraging the estimation theory lecture series and for facilitating the documentation that ultimately resulted in this book. A superficial attempt to apportion credit for the various ideas, concepts, and techniques appearing in this monograph would result in many injustices. At the risk of offending my fellow researchers in this area I have attempted, in Section I.4, to document the major historical contributions to this subject area. Since many of the key ideas were developed independently and/or with near simultaneity it is sometimes difficult to attribute due credits. Furthermore, the documentation is not encyclopedic and really covers only references that were familiar to the author or were recommended for inclusion by colleagues. Apologies are given to those fellow researchers whose works have been inadequately cited.

I would like to acknowledge my colleagues in JPL's Tracking and Orbit Determination Section, many of whom contributed to and/or motivated much of the material in this book. Particular credits are due to Messrs. C. Christensen, D. Curkendall, P. Dyer, J. Ellis, R. Hanson, F. Jordan, H. Lass, S. McReynolds, T. Nishimura, K. Rourke, and Ms. K. Thornton. Special thanks are due to Drs. J. McDanell and S. McReynolds for their reviews of the manuscript. Their criticisms and suggestions have greatly enhanced the clarity and scope of this work. Technical improvements due to Dr. McReynolds are appreciated and are here gratefully acknowledged.

I am also obliged to acknowledge the influence that Dr. P. Kaminski's Ph.D. thesis has had on computer implementation of square root information algorithms. His observation concerning the inclusion of process noise, one component at a time, contributed to the efficient computer implementation of these algorithms. Dr. Kaminski's major contribution, the square root information smoother, is discussed in Section I.4 and in Chapter X.

List of Symbols

Certain of the symbols defined here are used elsewhere in the text as temporary symbols; their appearance in such a context should cause no confusion.

Symbol	*Page*	*Remarks*
z	14	Observation m vector
$A(m, n)$	14	Observation coefficient matrix; sometimes subscripted (cf. p. 13)
J	14	Least squares performance functional; sometimes subscripted
$\|\| \ \|\|$	14	Euclidean length; in Section VI.4 this is a weighted Euclidean length (p. 125)
ls	14	Least squares subscript symbol
ν	14	Data equation error; generally normalized
$E(\cdot)$	15	Expectation operator
I	15	Identity matrix, sometimes subscripted
P	16	Covariance matrix; generally subscripted
P	139	Perturbation matrix
Λ	16	Information matrix; generally subscripted
$\tilde{}$	17	Tilde, used to signify *a priori*; also represents one-step prediction (p. 114)
$\hat{}$	17	Caret, used to represent optimal and filter results
R	17	Square root information matrix; often subscripted
T	18	Orthogonal transformation matrix
K	19	Gain
Tr	20	Matrix trace
U–D	24	Covariance matrix factors; U upper triangular and D diagonal
@	29	Symbol used to indicate a FORTRAN comment
**	31	FORTRAN symbol for exponentiation
PD	33	Positive definite abbreviation
$>$	34	"Greater than symbol"; to partially order PD matrices
ρ	34	Correlation coefficient

Symbol	*Page*	*Remarks*
e_j	35	jth coordinate basis vector
e_j	71	*A posteriori* measurement residual; usually subscripted
$\det(\cdot)$	36	Determinant symbol
L	38	Lower triangular matrix
$:=$	41	"Is replaced by" symbol
$\rightleftarrows$	43	"Is equivalent to" symbol
σ	48	Standard deviation (generally); often subscripted
$\perp$	59	Perpendicular (orthogonal) symbol
$\mathrm{sgn}(\cdot)$	63	Signum (sign) function
$K_j(\cdot)$	78	Notation for the components of a subscripted vector
$\tau_+, \tau_\times, \tau_\div, \tau_{\sqrt{}}$	84	UNIVAC 1108 arithmetic execution times
G	113	Process noise related matrix
Φ	113	State transition matrix; generally subscripted
w	113	White process noise; a notable exception occurs in Section VI.4
Q	114	Process noise covariance; generally diagonal
δ	114	Kronecker delta
k/N	114	Subscript notation denoting conditional estimation
$*$	114	Asterisk used to indicate fixed interval smoothing
$N(0, I)$	114	Statistical distribution that is unbiased and normalized
$\tilde{}, \hat{}, *$	115	Ornaments used to relate terms to one step prediction, filtering, and smoothing, respectively
p	135	Colored process noise vector, generally subscripted
y	135	Constant (bias) parameter vector
M	136	Colored noise transition matrix
c	139	Superscript used to denote "computed", *viz.* $x^{\mathrm{c}}, P^{\mathrm{c}}$
P_{con}	140	Consider covariance
S	144	Augmented column information array, generally subscripted
Sen	144	Sensitivity matrix; S is used on p. 139
$\gtrless$	163	Ordering symbols for positive definite matrices
P_ν^{a}	165	Actual measurement noise covariance; superscript a represents *actual*
y_{c}	165	Consider parameters; subscript c represents *consider*
$-$	167	Overbar; dummy symbol representing either $\tilde{}$ (prediction) or (filtering)
$\bar{P}_\nu^{\mathrm{a}}$	167	Prescaled actual measurement covariance
X	172	Augmented state model
$X^{y_{\mathrm{c}}}$	174	System response due to the y_{c} consider parameters
ΔX	175	State error; Δ represents a difference, i.e. actual minus estimated
$\mathrm{cov}(\cdot)$	178	Covariance function
SR	186	Square root of the data equation noise covariance
sr	192	Component submatrices of SR

I

Introduction

I.1 Introduction

The simplicity and versatility of the Kalman sequential filter [1–4] has established it as a fundamental tool in the analysis and solution of linear estimation problems. Numerical experience (cf. Schlee *et al.* (1967)[†]; Bellantoni and Dodge (1967); and Leondes [5, Chapter 14]) has shown that the Kalman filter algorithm is sensitive to computer roundoff and that numeric accuracy sometimes degrades to the point where the results cease to be meaningful. Thus estimation practitioners have been faced with a dilemma, i.e., how to take advantage of the simplicity and versatility of the Kalman filter without falling victim to its potential numeric inaccuracy and instability.

One pragmatic and frequently effective solution to this dilemma has been to identify and circumvent those estimation problems where numerical difficulties appear. For example, it is known that numerical difficulties have occurred in problems with perfect or nearly perfect measurements and with singular or nearly singular estimate covariance matrices. (These conditions are merely representative.) One approach to these problems has been to model them in such a way that the measurements and the dynamic model are sufficiently noisy to prevent near singularities from appearing.

[†]References enclosed in square brackets are listed at the end of the chapter and are generally annotated. References including the year of publication in parentheses are listed in the Bibliography at the end of the book. The distinction made is that references at chapter end are intended as collateral or supplementary reading; while material referred to in the Bibliography is primarily of historical or illustrative value. Chapter-end references are also included in the Bibliography.

An alternative and more consistently reliable solution to the numerical instability problem is to perform some or all of the computations using extra precision arithmetic. Now, however, new tradeoffs must be considered: i.e., introducing or artificially increasing noise covariance matrices to improve numeric stability versus the inaccuracies resulting from mismodeling, and the use of extra precision arithmetic to reduce the effects of roundoff errors versus an increase in computation and computer storage requirements.

An alternative to the introduction of mismodeling and/or "patching" (cf. Section V.5) of the Kalman filter algorithm via, sometimes ad hoc, stabilization techniques is to modify or replace the algorithm by one that is numerically better conditioned. The square root filter is such a solution, i.e., it has inherently better stability and numeric accuracy than does the Kalman filter. The improved numerical behavior of square root algorithms is due in large part to a reduction of the numerical ranges of the variables. Loosely speaking, one can say that computations which involve numbers ranging between 10^{-N} and 10^{N} are reduced to ranges between $10^{-N/2}$ and $10^{N/2}$. Thus the square root algorithms achieve accuracies that are comparable with a Kalman filter that uses twice their numerical precision.

Two different types of square root filters have been developed. One involves a square root of the error covariance matrix and corresponds to a factorization of the Kalman filter algorithm. The other involves a square root of the information matrix, and represents a fundamentally different approach to the estimation problem. These two types of square root algorithms are the principal subject of this monograph.

The utility and importance of the sequential square root filter has been demonstrated by Bellantoni and Dodge (1967), Schmidt (1972), Widnall (1973), and Christensen and Reinbold (1974). This monograph makes available in book form much of the material and experience that has heretofore been available only in archival publications and in reports with limited distribution. Our unified and self-contained exposition focuses on the computational aspects of linear least squares estimation with emphasis on square root estimation; it includes filter and smoother algorithm developments along with algorithms and techniques to systematically evaluate and analyze estimation performance.

I.2 Prerequisites

This monograph is primarily concerned with the development and discussion of various estimation algorithms. While this text is, with but a few exceptions, self-contained, it implicitly supposes a certain mathematical maturity and an at least casual familiarity with linear estimation. We also assume that our readers are interested in implementing estimation analysis algorithms on a digital computer; hence, a basic knowledge of FORTRAN is assumed.

We have chosen to restrict our analysis to linear estimation of discrete problems and have analyzed only the least squares and minimum variance criteria, because these criteria lead straightaway to the algorithms that are our prime concern. References [1–3] supply prerequisite or corequisite background material on probability and estimation theory. This monograph supplements those references because our work elaborates on topics that estimation texts either omit or cover only superficially. The cited texts contain other aspects and approaches to linear estimation as well as numerous examples and applications to which our algorithms can be applied.

Our mathematical prerequisites are modest and consist in the main of elementary matrix algebra such as is described by Noble [6]. We use this elementary background to develop those elements of numerical linear algebra that are essential for our estimation applications. This material is presented in Chapters III and IV. Those who are experienced with numerical solutions of discrete estimation problems may wonder at the absence of material on matrix ill-conditioning, the pseudoinverse, and the computation of approximate (rank deficient) matrix inverses. This material is important for the detection of numeric accuracy degradation and for its correction and/or compensation. We chose not to include these topics because they are already thoroughly covered in the recent text by Lawson and Hanson [7]. Furthermore, the square root estimation algorithms, which are the main topic of this book, are sufficiently reliable that numerical failure need not be a major concern for the vast majority of estimation applications (especially when process noise is present).

Our algorithms are intended for computer implementation. For this reason we have avoided compact or elegant representations and favored instead representations that translate directly into FORTRAN code. In addition, representative FORTRAN codes for the important algorithms are included for reference purposes. These codes require only a modest knowledge of FORTRAN IV, because they involve little more than knowledge of DO loops and IF statements.

I.3 Scope and Objectives

Existing texts on linear estimation quite adequately discuss the theory but are deficient in their discussion of the computational aspects of this subject. In recent years square root factorizations of the discrete Kalman filter and the use of orthogonal transformations has led to the creation of estimation algorithms that have reduced numerical sensitivity and are in many other ways superior to conventional filter mechanizations. One of the principal objectives of this book is to present these algorithms along with the mathematical background that is necessary for their understanding and utilization.

Square root algorithms have had difficulty gaining acceptance and, despite their demonstrated superior performance, have been received to a large extent as curiosities. That such an attitude is unwarranted becomes clear when one considers the various successful estimation applications that have relied on square root filtering techniques (viz. the Apollo lunar missions, the Mariner 9 Mars orbiter and the Mariner 10 Venus–Mercury space probe, and the precise aircraft navigation demonstrations described by Schmidt (1972) and Widnall (1973)). This cold reception is due perhaps to the fact that too few articles on square root estimation have appeared; and these have unfortunately led many to conclude that factorization techniques are too complicated (compared with, say, the Kalman filter), use too much computer storage, and involve too much computation. These conclusions are due in large part to a poor understanding of the factorization algorithms and the use of inefficient computer mechanizations. One of our aims is to demonstrate that this need not be the case, i.e., our factorization mechanizations can compete favorably with conventional algorithm mechanizations in terms of algorithm complexity, storage, and computer run times.

FORTRAN mechanization of the various estimation-related algorithms have been included so that the applications analyst can assess algorithm complexity at a glance and will at the same time be impressed with the paucity of computer code that is required. Inclusion of these FORTRAN codes is, of course, intended to increase the reference value of the monograph. FORTRAN codes for entire programs were not included because we want analysts in different fields to be free to tailor programs to their particular applications. Moreover, our experience is that computer programs tend to be too specialized or are complicated by generalized logic and programming minutiae. In both cases the simplicity of the algorithm mechanization is obscured.

Estimation formulae for the square root information filter (*SRIF*) are not well suited to hand computation, and effective computer implementation is heavily dependent on technology developed by numerical analysts, whose work is imperfectly understood by the estimation applications community. The result has been a misconception regarding the complexity and limitations of such estimation algorithms. One of the aims of this book is to dispel the misconception and to amalgamate estimation concepts with the requisite mathematical techniques of the numerical analyst.

One of the reasons that Kalman filtering continues to be endorsed by estimation practioners is that they are able to relate to it, i.e., the notions of filter gains and error covariance analysis have become old friends.

Analyzing the effects of using incorrect *a priori* statistics, unmodeled parameters, or suboptimal filter gains is thought by many to be carried out most simply and efficiently using covariance analysis. One of our objectives is to demonstrate that such analysis can be carried out equally well in terms of our "data equation" approach to estimation. Indeed, formulation of estimation problems in terms of data equations and square root information arrays leads directly to compact general purpose algorithms. The value of this approach is that additional insights into estimation phenomena are obtained, and at the same time one avoids the problem of detecting and compensating for Kalman filter numerical deterioration.

Our estimation algorithm developments correspond to both covariance and information matrix formulations. Each formulation has attributes to recommend it in particular situations. For example, covariance-related algorithms are best suited to applications that involve frequent estimates, scalar measurements, and (nearly) perfect *a priori* knowledge of model parameters. On the other hand, information-related algorithms are ideally suited to problems involving large amounts of data and which may also involve no *a priori* knowledge of model parameters. Our objective in developing these two kinds of algorithms and their corresponding attributes is to let the intended application dictate which methodology to apply.

I.4 Historical Perspectives

Deutsch (1965) and Sorenson (1970) give lucid accounts of least squares estimation from its beginnings with Gauss through the minimum variance state space approach of Kalman (1963). The sequential character of Kalman's solution to the linear parameter estimation problem made it ideal for implementation on a digital computer. Kalman's introduction of dynamic models that included process noise and his formulae for error covariance computation further contributed to the widespread acceptance of his ideas.

Because Kalman's covariance results do not depend on the actual data, analysts now quite routinely generate covariance matrices corresponding to candidate models and measurement strategies. In this way estimation accuracy is determined prior to the occurrence of the actual event. Indeed, covariance simulations have become an essential design tool that is used to determine the effects various variables have on estimation accuracy and to predict accuracy improvements that may result from the use of various kinds of measurements.

By introducing linearization about a nominal trajectory, Schmidt (cf. Smith *et al.* (1962)) demonstrated that Kalman's filter could be applied to nonlinear orbit determination problems. Instead of linearizing about the nominal trajectory, Schmidt proposed linearizing about the estimated trajectory. He reasoned that one could in this way better retain linearity of the approximation, because the esimated state is "on the average" closer to the actual state. This relinearized filter algorithm is called the *extended Kalman* filter (cf. Jazwinski (1970) and Sage and Melsa [3]). It is important to note that the extended Kalman filter covariance differs from the Kalman filter in that it depends on the estimate and thus cannot be evaluated off-line.

As researchers became more experienced with Kalman filtering they reported various divergence phenomena, i.e., numerical results were observed that contradicted predicted analytic behavior. Several types of divergence phenomena could be distinguished, viz.,

(a) divergence due to the use of incorrect *a priori* statistics and unmodeled parameters,
(b) divergence due to the presence of nonlinearities, and
(c) divergence due to the effects of computer roundoff.

Effects of type (a) were analyzed in progressively more detail and generality by Soong (1965), Nishimura (1966), Griffin and Sage (1969), and others. These authors provided covariance analyses of the errors introduced by using incorrect *a priori* information, and although they were aware that this can be a major cause of filter divergence, it was not until the work of Toda *et al.* (1969) that a significant demonstration of divergence due to *a priori* uncertainty was published.

Nonlinear problems of type (b) are generally unsolvable, although there are a few scattered exceptions. Application results to date have been, for the most part, obtained using approximation methods that are ad hoc. The results reported suggest, however, that although nonlinearity approximations have to be handled individually, the techniques are effective. Among the more successful of these methods is the iterated extended Kalman linearized filter (Jazwinski (1970)) which has already been mentioned, the Athans *et al.* (1968) second-order filter, and Jazwinski's J-adaptive filter (cf. Jazwinski (1970) and (1973)). We chose not to discuss nonlinear estimation techniques in this book because we believe that the pragmatic aspects of this subject have not significantly advanced beyond the material in Jazwinski's book.

Effects of type (c), divergence due to computer roundoff, have played a prominent role in the design of the various estimation algorithms discussed in this book. Difficulties relating to computer roundoff appeared in even the very early applications of Kalman filtering, although Schmidt's work (cf. Leondes [5, Chapter 13]) appears to contain the first documentation of this phenomenon. The effects of numerical errors are generally manifested in the appearance of computed covariance matrices that fail to retain nonnegativity (i.e., with nonnegative eigenvalues, cf. Chapter III). Among the methods that have been used to improve accuracy and to maintain nonnegativity and symmetry of the computed covariance are

(a) computation of only the upper (or lower) triangular entries, and then forcing symmetry (this also reduces the amount of computation),
(b) computation of the entire matrix and then averaging the off-diagonal elements,
(c) periodic testing and resetting of the diagonal elements (to retain nonnegativity) and the off-diagonal elements (to assure correlations that are less than unity),[†]
(d) replacement of the optimal Kalman measurement update[‡] $(I - KA)P$ by the more general expression $(I - KA)P(I - KA)^{\mathrm{T}} + KRK^{\mathrm{T}}$ (the greater numeric accuracy of this formulation is pointed out by Joseph (cf. Bucy and Joseph (1968)), and
(e) the use of larger process noise and measurement noise covariances.

One twist of idea (c) involves bounding the diagonal elements of the computed error covariance matrix, with (lower) bounds being chosen which reflect the engineer's confidence in the estimate. Another variation, this time of (e), is Schmidt's "ϵ" technique which overweights the most recent measurements (cf. Schmidt *et al.* (1968); and Leondes [5, Chapter 3]).

It is now established that methods involving square root matrices have numerical properties that are far superior to these ad hoc methods. The use of square root matrices implicitly preserves symmetry and assures non-negative eigenvalues for the computed covariance. Algorithms involving square root matrices have other desirable numerical attributes as well. Prominent among these is the reduced dynamic range of the numbers entering into the computations. This factor together with the greater accuracy that square root algorithms have been shown to possess (cf. Kaminski *et al.* (1971) and Kaminski's dissertation (1971)) directly affects computer word length requirements. A rule of thumb is that square root algorithms can use half the word length required by conventional non-square root algorithms.

Potter is credited with introducing square root factorization to the sequential estimation community. He showed (cf. Battin (1964)) that for scalar measurements it was possible to factor Kalman's covariance update formula and reformulate the estimation recursion so that the covariance

[†]In Chapter III we show, by example, that this method is unreliable.

[‡]K, A, P, and R are the Kalman gain, observation matrix, *a priori* state covariance, and measurement covariance, respectively.

matrix did not explicitly appear. Potter's covariance square root formulation was successfully demonstrated in the Apollo space program (cf. Leondes [5, Chapter 14]). However, Potter's formulation was restrictive because it did not include the effects of process noise and it applied only to the case of uncorrelated measurements. Andrews (1968) extended Potter's work to include process noise and also observed that Potter's updating algorithm could be applied to problems with correlated measurements, if first they were whitened (cf. Section III.4). Andrews' work was addressed to the continuous time problem, but it was not long thereafter that covariance square root solutions to the discrete time process noise problem were forthcoming (cf. Dyer and McReynolds (1969), Leondes [5, Chapter 3], and Kaminski *et al.* (1971)).

The gist of this discussion is that square root covariance filtering came about in response to Potter's desire to recast Kalman's sequential data processing algorithm in a numerically more stable form. Another square root approach to the parameter estimation problem, one due to Golub (1965), also emerged in the mid 1960s. Golub, a numerical analyst, was interested in accurate and stable methods for solving least square problems involving overdetermined systems of linear equations. Instead of modifying Kalman's solution, as Potter had done, Golub proposed using the numerically stable elementary orthogonal transformations that had been introduced by Householder (1958). Golub's least square solution enjoys two of the key properties of Kalman's solution, i.e., statistical interpretation of results and sequential processing of information. His contribution appears to have been obscured, however, perhaps for the following reasons: his work appeared in a numerical analysis journal which was unfortunately, not widely read by estimation engineers; his proposed matrix triangularizations using orthogonal transformations seemed too complicated and this appeared to detract from the practicality of his method; and his method did not appear capable of generalization to include process noise effects. These objections and others of a similar nature have gradually been put to rest.

Hanson and Lawson (1969) elaborated on the algorithmic details of Golub's triangularization procedure. Their work forms the basis of the Jet Propulsion Laboratory's (JPL) least squares ephemeris correction program, and also played a key part in the Mariner 9 1971 Mars space mission. The significance of their work is that it demonstrated the efficacy and reliability of Householder orthogonalization for large-scale, poorly conditioned estimation problems. The experience of JPL personnel with the Hanson–Lawson estimation software revealed that the square root information array, a matrix that only implicitly contains the estimate and the estimate error covariance, is in fact a natural vehicle with which to derive and implement estimation formulae.

Cox (1964) pointed out that a quadratic performance functional could be used to include the effects of process noise. From this work it was clear how one could extend the square root information array to include process noise effects (cf. Dyer and McReynolds (1969)). The Dyer–McReynolds filter algorithm was used with notable success for the navigation computations on the Mariner 10 1973 Venus–Mercury space mission (cf. Christensen and Reinbold (1974)). This success established the reliability and utility of the sequential square root information filter for critical real-time mission operations.

One of the unexpected benefits accruing to the Dyer–McReynolds approach to estimation was that smoothed, postflight estimates and associated covariances could be obtained with a very modest increase of computation (cf. Kaminski's dissertation (1971); Kaminski and Bryson (1972); and Bierman (1974)). Indeed, Bierman's work showed that of the various smoothing formulations that have appeared in the literature the formulation resulting from the Dyer–McReynolds information array is the most efficient.

The Kaminski *et al.* (1971) survey on square root filtering together with the successful applications[†] of this technology stimulated interest in this subject. One stumbling block that remained to hamper acceptance of the square root filter was its lack of intuitive statistical appeal. This obstacle was overcome for the SRIF by Bierman (1974) with his "*data equation*" approach to estimation, and for the covariance square root by Morf and Kailath (1974) who viewed the square root process as a Gram–Schmidt orthogonalization of an appropriately augmented state vector. These statistical interpretations are certain to result in new algorithms and engineering insights.

1.5 Chapter Synopses

Our quasitutorial, expository presentation highlights those computational aspects of discrete linear estimation that lead to efficient, flexible, and insightful analyses. Key topics to be discussed include the Kalman

[†]In addition to the JPL applications and extensions of the Dyer–McReynolds work, there are avionics applications of square root covariance filtering by Schmidt (1972) and Carlson (cf. Widnall (1973)).

filter, the Potter and Carlson square root covariance measurement updating formulae, a square root free triangular factorization method for measurement updating, the Dyer–McReynolds square root information filter, covariance performance analysis, considering the effects of unestimated parameters in both the Kalman and square root information filters, sensitivity computations, and discrete fixed interval smoothing.

Chapter II is a review of the classical least squares problem. Recursive solutions of the Gauss normal equations are not only of historical interest, they are still used to solve problems in orbit determination, satellite navigation, missile tracking, economics, etc. Because of its prominent role in modern estimation we include a description of the Kalman filter data processing algorithm. It is ideally suited for real-time updating of the least squares estimate and estimate error covariance and is the starting point for the development of covariance square root algorithms. Much of our SRIF development which is discussed in later chapters was motivated by, and will be compared with, the Kalman filter. The chapter concludes with Potter's square root mechanization of the Kalman algorithm. FORTRAN mechanizations of the Kalman and Potter algorithms are included as appendices.

Chapter III contains a review of various utilitarian properties of positive definite (PD) matrices that are useful for covariance analysis and are requisite for our development of square root estimation. FORTRAN mechanizations of several important PD algorithms are included as appendices. In Chapter IV the Householder orthogonal transformation is introduced. Our SRIF development exploits the properties and algorithms that are described in this chapter. In Chapter V the orthogonal transformation material is used to develop a square root analogue of the normal equation recursions (cf. Chapter II). This result, due to Golub (1965), is the data processing portion of the SRIF. In this chapter we also introduce an alternative data processing algorithm based on a triangular square root free factorization of the error covariance matrix. Our U–D algorithm (U and D being the upper triangular and diagonal factors, respectively) is a stable mechanization of a decomposition proposed by Agee and Turner (1972). The algorithm for $UD^{1/2}$ corresponds to the upper triangular covariance square root algorithm proposed by Carlson (1973). Computation count comparisons are made of the Potter, Golub, and Carlson square root algorithms, the U–D factorization, and the Kalman filter. A FORTRAN mechanization of the U–D algorithm is included as an appendix.

In Chapter VI the propagation portion of the SRIF, the Kalman filter, and the square root covariance filter are developed. These algorithms solve the prediction problem. This chapter also contains a discussion of the duality between covariance and information formulations of the filtering problem.

Chapter VII is an elaboration of the SRIF material covered in the previous two chapters. Computer implementation of the SRIF is expedited when the model is partitioned to distinguish the colored noise and bias parameters. Advantages accruing to such a partitioning are that computer storage requirements are significantly reduced and the special properties of these variables result in additional analytic perceptions. In particular, the notions of sensitivity analysis, variable order filtering, and *consider covariance*† analysis carry over unchanged from those analyses that did not involve process noise. This chapter also contains a colored noise processing algorithm that permits a unified treatment of colored and white noise. FORTRAN mechanizations of the various notions of this chapter are presented as an appendix. This computer mechanization, used to design the filter portion of JPL's Orbit Determination Program, involves only a fraction of the storage and computation used in earlier SRIF computer implementations.

Chapters VIII and IX are devoted to the accuracy analysis and performance prediction aspects of filtering. Recursive algorithms are derived by analyzing the effects of incorrect *a priori* statistics and unmodeled parameters. Chapter X focuses on smoothing, i.e., utilization of past and current measurement information to estimate the model's past history. Smoothing is important because many estimation applications do not involve (instantaneous) real-time decision making. Interesting and important features of our smoothing algorithm mechanization, which is an addendum to the SRIF, are that it requires only a modicum of additional computation and its structure is such that it preserves those features of the SRIF associated with sensitivity analysis, variable order estimation, and the consider covariance. In this chapter on smoothing we also review the covariance-related smoothing algorithms due to Rauch *et al.* (1965) along with Bierman's (1973a) modification of the Bryson–Frazier (1963) smoother.

References

[1] Bryson, A. E., and Ho, Y. C., "Applied Optimal Control." Ginn (Blaisdell), Waltham, Massachusetts (1969).

†Consider covariances are error covariance contributions due to the effects of unmodeled bias parameters. The term "consider" comes about because one speaks of considering the effects of unestimated parameters.

[2] Meditch, J. S., "Stochastic Optimal Linear Estimation and Control." McGraw-Hill, New York (1969).

[3] Sage, A. P., and Melsa, J. L., "Estimation Theory with Applications to Communications and Control." McGraw-Hill, New York (1970).

[4] Sorenson, H. W., Least squares estimation: from Gauss to Kalman. *IEEE Spectrum* **7**, 63–68 (1970).

[5] Leondes, C. T. (ed.), Theory and applications of Kalman filtering, *NATO Advisory Group for Aerospace Res. and Develop.*, AGARDograph 139 (1970).

[6] Noble, B., "Applied Linear Algebra." Prentice-Hall, Englewood Cliffs, New Jersey (1969).

[7] Lawson, C. L., and Hanson, R. J., "Solving Least Squares Problems." Prentice-Hall, Englewood Cliffs, New Jersey (1974).

II

Review of Least Squares Data Processing and the Kalman Filter Algorithm

II.1 Introduction

We begin our investigation of sequential least squares estimation with a review of the "classical" least squares problem, by which we mean the problem of determining parameter estimates from a system of linear equations. Analyzing the simple parameter estimation problem will prepare us for problems that involve stochastic variables. The results of the parameter estimation problem are important for pedagogical reasons and because "classical" parameter estimation has enjoyed great popularity among orbit determination practitioners, economists, and statisticians. By relating classical parameter estimation techniques with the modern method of square root information estimation we hope to encourage greater utilization and understanding of this latter approach.

In recent years a different estimation procedure, recursive minimum variance estimation (popularized by Kalman), has come into wide use. We shall develop the Kalman recursions and will discuss a square root formulation of the Kalman algorithm, due to Potter [2], which is more accurate and stable. FORTRAN mechanizations that highlight the similarities of the two algorithms are given in Appendices II.D and II.E. In Chapter V other parameter estimation algorithms will be developed and discussed. There we will compare the respective advantages, limitations, and numerical efficiencies of the various methods.

II.2 Linear Least Squares

Suppose that we are given the linear system

$$z = Ax + \nu \tag{2.1}$$

where z is an m vector of observations, x is the n vector of variables that are to be estimated, $A(m, n)$ is the coefficient matrix† of "*observation partials*," and ν is an m vector of observation errors.

The least squares solution x_{ls} to system (2.1) is chosen to minimize the mean square observation error:

$$J = \sum_{j=1}^{m} \nu(j)^2 = \nu^T\nu$$

i.e.,

$$J(x) = (z - Ax)^T(z - Ax) \tag{2.2}$$

Remark: When more than one x minimizes the mean square error J, we choose x_{ls} as the minimizing x of smallest Euclidean length

$$\|x\| = \left(\sum_{j=1}^{n} x(j)^2\right)^{1/2}$$

One can show that this uniquely defines x_{ls}.

Thus $J(x)$ is nonnegative and quadratic in the components of x so that a necessary and sufficient condition for a minimum is that the first variation, $\delta J(x)$, vanishes.

Remark: The first variation can be defined via the abbreviated expansion

$$J(x + \delta x) - J(x) = \delta J(x) + o(\delta x)$$

where $o(\delta x)$ is such that

$$\lim_{\|\delta x\| \downarrow 0} (o(\delta x)/\|\delta x\|) = 0$$

†We adopt the FORTRAN convention of dimensioning matrices, viz., $A(m, n)$ is a matrix with "m" rows and "n" columns. Although $A(m, n)$ is also used to describe an element of the matrix A there is no danger of confusion; the context will make the meaning clear.

$\delta J(x)$ can be obtained by formally applying the rules of differentiation, noting that since these are matrix operations they do not commute. Thus from (2.2) we obtain

$$\begin{aligned}\delta J(x) &= \delta\left[(z - Ax)^{\mathrm{T}}\right](z - Ax) + (z - Ax)^{\mathrm{T}}\,\delta[z - Ax] \\ &= (-A\,\delta x)^{\mathrm{T}}(z - Ax) + (z - Ax)^{\mathrm{T}}(-A\,\delta x) \\ &= \delta x^{\mathrm{T}}(A^{\mathrm{T}}Ax - A^{\mathrm{T}}z) + \left[\delta x^{\mathrm{T}}\,(A^{\mathrm{T}}Ax - A^{\mathrm{T}}z)\right]^{\mathrm{T}}\end{aligned}$$

For $\delta J(x)$ to vanish for all variations δx, it is necessary that x satisfy the *normal equations*

$$A^{\mathrm{T}}Ax = A^{\mathrm{T}}z \tag{2.3}$$

Remark: When $m > n$ and A has rank n, then $A^{\mathrm{T}}A$ is (theoretically) nonsingular (see Appendix II.A) and the solution x_{ls} is unique, i.e.,

$$x_{\mathrm{ls}} = (A^{\mathrm{T}}A)^{-1}A^{\mathrm{T}}z \tag{2.4}$$

It will generally be assumed in what follows that A has full rank.

II.3 Statistical Interpretation of the Least Squares Solution

Suppose that the observation errors ν of Eq. (2.1) are random variables with known (measurement) covariance. We assume the equations have been normalized† so that

$$E(\nu) = 0 \qquad \text{and} \qquad E(\nu\nu^{\mathrm{T}}) = I_m \tag{3.1‡}$$

with I_m being the m dimensional identity. Under these assumptions we find, from the normal equations (2.3), that

$$A^{\mathrm{T}}Ax_{\mathrm{ls}} = A^{\mathrm{T}}z = A^{\mathrm{T}}Ax + A^{\mathrm{T}}\nu$$

†The assumption of a normalized measurement covariance simplifies many of our formulae, and except where noted, will be assumed throughout this work. Generality is not compromised because one can return to the unnormalized measurement by replacing z and A by $S_\nu z$ and $S_\nu A$, where S_ν is a square root of P_ν, the covariance of ν.

‡The basic property of the expectation operator E that we rely on is linearity, i.e., $E(AxB + CyD) = AE(x)B + CE(y)D$, where matrices A, B, C, and D are deterministic and x and y are random matrices.

so that

$$A^{\mathrm{T}}A(x_{\mathrm{ls}} - x) = A^{\mathrm{T}}\nu \tag{3.2}$$

From this expression it is evident that

$$x = E(x_{\mathrm{ls}}) \tag{3.3}$$

and

$$(A^{\mathrm{T}}A)E\left[(x_{\mathrm{ls}} - x)(x_{\mathrm{ls}} - x)^{\mathrm{T}}\right](A^{\mathrm{T}}A) = A^{\mathrm{T}}E(\nu\nu^{\mathrm{T}})A,$$

which simplifies to

$$P_{x_{\mathrm{ls}}} \overset{\Delta}{=} E\left[(x - x_{\mathrm{ls}})(x - x_{\mathrm{ls}})^{\mathrm{T}}\right] = (A^{\mathrm{T}}A)^{-1} \tag{3.4}$$

Thus $(A^{\mathrm{T}}A)^{-1}$ is the estimate error covariance, and $P_{x_{\mathrm{ls}}}^{-1}$, called the *information matrix*, is $A^{\mathrm{T}}A$. Notation for the information matrix will be Λ_x.

Remark: Equation (3.2) shows that x_{ls} is an unbiased estimate.

II.4 Inclusion of a Priori Statistics

Suppose that in addition to the linear system (2.1), we have an *a priori* unbiased estimate of x, $\tilde{x}$ and an *a priori* information matrix $\tilde{\Lambda}$.†

One can include this information in the least squares context by considering the modified performance functional $J_1(x)$,

$$J_1(x) = (x - \tilde{x})^{\mathrm{T}}\tilde{\Lambda}(x - \tilde{x}) + (z - Ax)^{\mathrm{T}}(z - Ax) \tag{4.1}$$

$$= x^{\mathrm{T}}(\tilde{\Lambda} + A^{\mathrm{T}}A)x - x^{\mathrm{T}}(\tilde{\Lambda}\tilde{x} + A^{\mathrm{T}}z) - \left[x^{\mathrm{T}}(\tilde{\Lambda}\tilde{x} + A^{\mathrm{T}}z)\right]^{\mathrm{T}}$$

$$+ \tilde{x}^{\mathrm{T}}\tilde{\Lambda}\tilde{x} + z^{\mathrm{T}}z \tag{4.2}$$

Comparing (4.2) with $J(x)$ expanded, it is easy to conclude that the minimizing argument of J_1, $\hat{x}_{\mathrm{ls}}$, must satisfy

$$(\tilde{\Lambda} + A^{\mathrm{T}}A)\hat{x}_{\mathrm{ls}} = \tilde{\Lambda}\tilde{x} + A^{\mathrm{T}}z \tag{4.3}$$

†The reader is cautioned that in the estimation literature the ˜ symbol is sometimes used to denote estimation errors.

A modest modification of our earlier arguments gives a statistical interpretation to $\hat{x}_{ls}$, i.e.,

$$\begin{aligned}\tilde{\Lambda}\tilde{x} + A^{T}z &= \tilde{\Lambda}x + \tilde{\Lambda}(\tilde{x} - x) + A^{T}Ax + A^{T}\nu \\ &= (\tilde{\Lambda} + A^{T}A)x + \left[A^{T}\nu + \tilde{\Lambda}(\tilde{x} - x)\right] \\ &= (\tilde{\Lambda} + A^{T}A)\hat{x}_{ls}\end{aligned}$$

Thus

$$(\tilde{\Lambda} + A^{T}A)(x - \hat{x}_{ls}) = \tilde{\Lambda}(x - \tilde{x}) - A^{T}\nu \tag{4.4}$$

and since $x - \tilde{x}$ and ν are zero mean, this implies that

$$x = E(\hat{x}_{ls}) \tag{4.5}$$

i.e., $\hat{x}_{ls}$ is also an unbiased estimate.

Assuming that $E[(x - \tilde{x})\nu^{T}] = 0$ (that the observation noise and the *a priori* estimate errors are independent), one can "square up" Eq. (4.4), i.e., multiply the equation by its transpose, and obtain

$$P_{\hat{x}_{ls}} \triangleq E\left[(x - \hat{x}_{ls})(x - \hat{x}_{ls})^{T}\right] = (\tilde{\Lambda} + A^{T}A)^{-1} \tag{4.6}$$

Remark: $\tilde{\Lambda}$ need not be nonsingular. Indeed, in many problems of interest certain components of x are completely unknown so that corresponding rows and columns of $\tilde{\Lambda}$ are filled with zeros. Observe that (4.3) and (4.6) for $\hat{x}_{ls}$ and $P_{\hat{x}_{ls}}$ reduce to the previous results, (2.3) and (3.4), when $\tilde{\Lambda} = 0$, i.e., when there is no *a priori* information.

II.5 Recursions for the Least Squares Information Processor

Let us suppose that the matrix $\tilde{\Lambda}$ is factored as

$$\tilde{\Lambda} = \tilde{R}^{T}\tilde{R} \tag{5.1}$$

An Aside: Factorization of a symmetric matrix with nonnegative eigenvalues (a positive semidefinite matrix) is always possible, and $\tilde{R}^{T}$ is called a *square root* of $\tilde{\Lambda}$. The notion of a square root matrix is a generalization of the square root of a number. As one might expect, square roots are nonunique. They are, however, related to one another by orthogonal transformations (i.e., if S_1 and S_2 are square roots of positive semidefinite matrix P; $P = S_1S_1^{\mathrm{T}} = S_2S_2^{\mathrm{T}}$; then there exists a matrix T such that $S_2 = S_1T$ and $TT^{\mathrm{T}} = T^{\mathrm{T}}T = I$). By restricting ourselves to positive semidefinite matrices we are able to avoid the use of complex arithmetic.

Square root matrices are very important for statistical analysis. They are useful both in theory and for algorithm implementation. Indeed, the use of square root matrices in linear estimation is one of the principal topics of this book. In Chapters III and IV we shall discuss the construction of square root matrices. ■

The purpose of the square root factorization of Λ is that it allows us to write

$$(\tilde{x} - x)^{\mathrm{T}}\tilde{\Lambda}(\tilde{x} - x) = (\tilde{z} - \tilde{R}x)^{\mathrm{T}}(\tilde{z} - \tilde{R}x) \tag{5.2}$$

where

$$\tilde{z} = \tilde{R}\tilde{x} \tag{5.3}$$

Comparing this with the least squares performance functional $J(x)$ leads to the important observation that *a priori estimate–covariance information can be interpreted as additional observations*. Thus $J_1(x)$ can be interpreted as simply applying the least squares performance functional to the augmented system

$$\begin{bmatrix} \tilde{z} \\ z \end{bmatrix} = \begin{bmatrix} \tilde{R} \\ A \end{bmatrix} x + \begin{bmatrix} \tilde{\nu} \\ \nu \end{bmatrix} \tag{5.4}$$

Substituting the partitioned matrices into (2.3) and (3.4) leads to the estimate–covariance recursions (4.3) and (4.6).

Remarks: (1) The results are recursive because the current estimate and covariance become the *a priori* and are combined with the new data to form an updated estimate and covariance.

(2) The interpretation of the *a priori* as additional observations is the cornerstone of the square root information filter.

(3) It should be noted that the estimate–covariance results, (4.3) and (4.6), do not depend on which square root was used. Use of different square root matrices does not (theoretically) alter the results. This freedom will later be exploited to select square root matrices that are simple to compute and that enhance numerical accuracy.

(4) Recursive calculation using (4.3) suggests computation of an estimate at each step. If only the final estimate is desired it is best not to process data sequentially using (4.3), but instead to accumulate $\Sigma A_j^{\mathrm{T}}A_j$

and $\Sigma A_j^{\mathrm{T}} z_j$, and then to use the estimate and covariance formulae for the final step (see Appendix II.C).

II.6 Kalman Filter Data Processing

The computational efficiency of Eqs. (4.3) and (4.6) diminishes when the dimension of the observation vector m is less than n. The reason for this is that these equations imply the inversion of a matrix of dimension n, and it is possible to reformulate the result involving the inverse of an $m \times m$ matrix. Using the matrix inversion lemma of Appendix II.B and some matrix algebra, one can derive the Kalman data processing equations (Appendix II.D). We choose to derive these results using a somewhat more enlightening approach.

The result, Eq. (4.3), shows that the least squares estimate† $\hat{x}$ is a linear combination of the *a priori* estimate $\tilde{x}$ and the new data,

$$z = Ax + \nu$$

i.e.,

$$\hat{x} = L\tilde{x} + Kz \tag{6.1}$$

Motivated by Eqs. (3.3) and (4.5) we require $\hat{x}$ to be unbiased, i.e.,

$$x = E(\hat{x}) = LE(\tilde{x}) + KE(z) = (L + KA)x \tag{6.2}$$

Therefore,

$$I = L + KA \tag{6.3}$$

and

$$\hat{x} = \tilde{x} + K(z - A\tilde{x}) \tag{6.4}$$

Since we are speaking of random variables (cf. (3.1)), it is natural to replace the performance functional J_1, Eq. (4.1), with a probabalistic one, i.e.,

$$J_2(K) = E\left(\sum_{i=1}^{n} [\hat{x}(i) - x(i)]^2\right) = E\left[(\hat{x} - x)^{\mathrm{T}}(\hat{x} - x)\right] \tag{6.5}$$

Remark: $\hat{x}$ depends on K because of (6.4), and that is why we write $J_2(K)$. Also, the choice of (6.5) gives rise to the name *minimum variance*.

†In this section we write $\hat{x}$ instead of x_{ls}.

Let us express J_2 in terms of covariances

$$J_2(K) = E\{\mathrm{Tr}[(\hat{x} - x)^{\mathrm{T}}(\hat{x} - x)]\} \quad \text{(the trace of a scalar is the scalar)}$$

$$= E\{\mathrm{Tr}((\hat{x} - x)(\hat{x} - x)^{\mathrm{T}})\} \quad (\mathrm{Tr}(AB) = \mathrm{Tr}(BA))$$

$$= \mathrm{Tr}(E\{(\hat{x} - x)(\hat{x} - x)^{\mathrm{T}}\})$$

(linearity of the expectation operation)

Therefore

$$J_2(K) = \mathrm{Tr}(\hat{P}) \tag{6.6}$$

where $\hat{P}$ is the *a posteriori* covariance, i.e., the error covariance of the estimate based on the *a priori* information and the data. Thus minimizing J_2 corresponds to selecting the estimate to minimize the *a posteriori* error covariance trace.

Remark: The trace of a square matrix (the algebraic sum of its diagonal elements) appears quite often in estimation and control problems. Derivations of many frequently occurring trace functions can be found in [6], and [7] contains a summary of important trace relationships.

To continue further in the development of a K which minimizes $J_2(K)$, we return to (6.4) and write

$$\hat{x} - x = \tilde{x} - x + K(Ax + \nu - A\tilde{x})$$

$$= (I - KA)(\tilde{x} - x) + K\nu \tag{6.7}$$

so that

$$\hat{P} = E\{(\hat{x} - x)(\hat{x} - x)^{\mathrm{T}}\}$$

$$= E\{[(I - KA)(\tilde{x} - x) + K\nu][(\tilde{x} - x)^{\mathrm{T}}(I - KA)^{\mathrm{T}} + \nu^{\mathrm{T}}K^{\mathrm{T}}]\}$$

$$= (I - KA)\tilde{P}(I - KA)^{\mathrm{T}} + KK^{\mathrm{T}} \tag{6.8}$$

$$= K(A\tilde{P}A^{\mathrm{T}} + I)K^{\mathrm{T}} - KA\tilde{P} - \tilde{P}A^{\mathrm{T}}K^{\mathrm{T}} + \tilde{P}$$

$$= (K - \tilde{P}A^{\mathrm{T}}D^{-1})D(K - \tilde{P}A^{\mathrm{T}}D^{-1})^{\mathrm{T}}$$

$$+ \tilde{P} - \tilde{P}A^{\mathrm{T}}D^{-1}A\tilde{P} \tag{6.9}$$

where

$$D = A\tilde{P}A^{\mathrm{T}} + I_m \tag{6.10}$$

Substituting (6.9) into performance functional (6.6), and taking the first variation, we obtain

$$\begin{aligned}\delta J_2(K) &= \mathrm{Tr}\left\{\delta K(DK^{\mathrm{T}} - A\tilde{P}) + \left[\delta K(DK^{\mathrm{T}} - A\tilde{P})\right]^{\mathrm{T}}\right\} \\ &= 2\,\mathrm{Tr}\left\{\delta K(DK^{\mathrm{T}} - A\tilde{P})\right\}\end{aligned}$$

Therefore, the first variation vanishes provided

$$DK^{\mathrm{T}} = A\tilde{P} \tag{6.11}$$

or

$$K = \tilde{P}A^{\mathrm{T}}D^{-1} = \tilde{P}A^{\mathrm{T}}\left(A\tilde{P}A^{\mathrm{T}} + I_m\right)^{-1} \tag{6.12}$$

To obtain the covariance of the estimate, substitute (6.11) into (6.9) and obtain

$$\hat{P} = \tilde{P} - KA\tilde{P} = \tilde{P} - KDK^{\mathrm{T}} \tag{6.13}$$

Remarks: (1) It is easy to show that K given by (6.12) is the same as the result implied by (4.3), viz., $K = (\tilde{P}^{-1} + A^{\mathrm{T}}A)^{-1}A^{\mathrm{T}}$. Thus minimizing the *a posteriori* squared error is the same as minimizing the *a posteriori* covariance trace, and this is equivalent to minimizing the measurement residual† *a priori* weighted deterministic least squares problem.

(2) The algorithm starts with an estimate and covariance and computes an updated estimate and covariance. The information processor computes estimates and covariances only for output purposes.

(3) This algorithm is most efficiently and simply mechanized by processing vector measurements one component at a time. In this case the equations can be arranged (cf. Appendix II.D) to make maximal use of vector inner and outer products as well as to reduce computer storage requirements.

(4) By its construction the Kalman estimate is minimum variance, unbiased, and linear.

(5) The matrix K is called the *Kalman gain*; it weights the predicted

†In the literature one usually sees the minimum variance criterion compared with a weighted sum of residual errors. Although it may be more subtle, we too are doing this because our measurements are normalized to unit variance; and thus our residuals are also weighted.

residual $\tilde{\nu} = z - A\tilde{x}$. The matrix D is the covariance of the predicted residual, i.e., $D = E(\tilde{\nu}\tilde{\nu}^{\mathrm{T}})$.

II.7 Potter's Mechanization of the Kalman Algorithm

Numerical experience with the Kalman algorithm, Eq. (6.13), indicates that it is sensitive to the effects of computer roundoff and is susceptible to an accuracy degradation due to the differencing of positive terms.† Potter [2] recognized that numerical accuracy degradation is often accompanied by a computed covariance matrix that loses its positive definiteness (cf. Chapter III). By reformulating the Kalman algorithm in terms of a square root covariance matrix he thought to preserve nonnegativity of the computed covariance (which is only an implicit quantity in his algorithm). It turns out that introduction of the covariance square root does indeed improve numeric accuracy and stability. When using Potter's algorithm, however, one can no longer use the nonnegativity property of the covariance matrix to detect the onset of numeric degeneracy. Thus accuracy degradation, when it occurs, is harder to detect.

An Aside: One test for algorithm numeric sensitivity is to redundantly solve the same problem using permuted lists of variables. A more reliable, but off-line, technique for determining whether in a given problem there is numeric accuracy deterioration due to computer roundoff is to solve the problem twice using single and double precision arithmetic. (These tests are suggested in Wampler (1970).) A subtlety to be aware of when making the single precision/double precision comparison is that one must be careful to use only single precision input values; otherwise we may only be exposing a numeric sensitivity of the problem and not of the algorithm. Examples illustrating algorithm numeric deterioration are discussed in Section V.5. ■

To obtain the Potter mechanization of the Kalman algorithm we set

$$\tilde{P} = \tilde{S}\tilde{S}^{\mathrm{T}}, \qquad \hat{P} = \hat{S}\hat{S}^{\mathrm{T}}$$

and rearrange (6.13) as

$$\begin{aligned} \hat{P} = \hat{S}\hat{S}^{\mathrm{T}} &= \tilde{S}\tilde{S}^{\mathrm{T}} - \tilde{S}\tilde{S}^{\mathrm{T}}A^{\mathrm{T}}D^{-1}A\tilde{S}\tilde{S}^{\mathrm{T}} \\ &= \tilde{S}\,[I - vD^{-1}v^{\mathrm{T}}]\tilde{S}^{\mathrm{T}} \end{aligned} \tag{7.1}$$

†For example, suppose that two numbers that agree to four digits are differenced. If the original numbers were only known to, say, six-digit accuracy, then the difference will have only two-digit accuracy.

where

$$v = (A\tilde{S})^{\mathrm{T}} \qquad \text{and} \qquad D = v^{\mathrm{T}}v + I$$

From (7.1) it is easy to conclude that one could define

$$\hat{S} = \tilde{S}(I - vD^{-1}v^{\mathrm{T}})^{1/2} \tag{7.2}$$

Potter noted that when the measurement was a *scalar*, $I - vD^{-1}v$ could be factored as

$$I - D^{-1}vv^{\mathrm{T}} = (I - \alpha vv^{\mathrm{T}})^2 \tag{7.3}$$

where α is a root of the quadratic equation

$$(v^{\mathrm{T}}v)\alpha^2 - 2\alpha + D^{-1} = 0 \tag{7.4}$$

Remark: This factorization is a key ingredient of Potter's algorithm. It can, however, be traced back to much earlier work. Householder [3] calls such matrices *elementary* and defines them as $\overline{E}(u, v; \sigma) = I - \sigma uv^{\mathrm{T}}$, where σ is a scalar and u and v are vectors. He points out that

$$\overline{E}(u, v; \sigma)\overline{E}(u, v; \alpha) = \overline{E}(u, v; \sigma + \alpha - \sigma\alpha v^{\mathrm{T}}u)$$

Evidently taking $u = v$ and $\sigma = \alpha$ corresponds to (7.3) and (7.4). Householder mentions that these elementary matrices go back at least to Turnbull and Aitken (1930) and Wedderburn (1934). Our observation at this point should in no way detract from the credit due Potter because it is his application of the result that is important.

The roots of (7.3) are $\alpha = 1/\left[\sqrt{D}\,(\sqrt{D} \pm 1)\right]$, and to avoid cancellation problems we choose

$$\alpha = 1/\left[\sqrt{D}\,(\sqrt{D} + 1)\right] \tag{7.5}$$

When (7.3) is substituted into (7.2) one obtains

$$\hat{S} = \tilde{S} - \left[\sqrt{D}\,/(\sqrt{D} + 1)\right]Kv^{\mathrm{T}}; \qquad K = Sv/D \tag{7.6}$$

Equation (7.6) is the key ingredient for Potter's recursive data processing algorithm. The algorithm is obtained from Eqs. (6.4), (6.12), and (7.6); a mechanization and FORTRAN realization are given in Appendix II.E.

Unlike the covariance matrix, $\hat{S}$ contains n^2 distinct elements. One could reduce computer storage requirements and transform $\hat{S}$ to a triangular form that involved fewer elements. The computation of a triangular covariance square root, however, involves n scalar square roots instead of the one appearing in Eq. (7.6). There is a concern about this because computation of scalar square roots is very time consuming (compared with other arithmetic operations); and this can, under certain conditions,† result in a very expensive algorithm. On the other hand, Carlson [4] has constructed a triangular covariance square root algorithm and has shown that, except for extreme cases, its computational requirements are not excessive. Carlson's work will be discussed in Chapter V.

It is of interest to note that one can avoid involving any square roots if one uses a covariance factorization of the form $P = UDU^{\mathrm{T}}$, where U is triangular and D diagonal. Agee and Turner (1972) seem to be the first to have combined this factorization with the Kalman filtering problem. In Chapter V we will modify Carlson's development to accommodate this factorization and will derive U–D recursions that are numerically stable and very nearly as efficient as the conventional, but numerically unstable, Kalman covariance update (6.13).

II.8 Computational Considerations Associated with Covariance Data Processing

If the Kalman algorithm is to be used [Eqs. (6.4), (6.12), and (6.13)], one should process the observations one at a time. This reduces computation because vector operations replace matrix operations and storage requirements can be reduced by having the *a posteriori* estimate and covariance displace the *a priori* estimate and covariance. This point is illustrated in the FORTRAN mechanizations appearing in Appendices II.D and II.E. We further note that Eq. (6.8) should be used to update the covariance‡ instead of (6.13), because the latter is known to contribute numerical instabilities in various poorly conditioned systems (such as those occurring, for example, in orbit determination). When the Potter mechanization is used there seems to be no need for the introduction of numerical stabilization.

Appendices II.C–II.E present the information data processing, Kalman data processing, and Potter algorithms that were discussed in this chapter. It is interesting to note that although the Kalman algorithm and the Potter mechanization appear quite similar, the latter algorithm enjoys signifi-

†Viz., when there are many measurements of a large-dimensioned parameter vector.

‡A more stable form of this result, which was determined empirically, is given in Appendix II.E.

cantly greater numeric stability. This is somewhat surprising since both algorithms have essentially equal computational requirements (cf. Section V.4). Thus, of the two, the Potter method is the better data processing algorithm.

APPENDIX II.A Proof that an Overdetermined System with Full Rank Has a Nonsingular Normal Matrix

A necessary condition for a minimum to the least squares problem

$$J = \|Ax - z\|^2$$

is that x satisfy the normal equations

$$A^{\mathrm{T}}Ax = A^{\mathrm{T}}z, \qquad A(m, n)$$

We now demonstrate that if $m \geqslant n$ and A has full rank n, then the normal matrix is nonsingular.

Let us write

$$A = \begin{bmatrix} a_1^{\mathrm{T}} \\ \vdots \\ a_m^{\mathrm{T}} \end{bmatrix}$$

where the a_j are n vectors. Then

$$A^{\mathrm{T}}A = \sum_{j=1}^{m} a_j a_j^{\mathrm{T}} \tag{A.1}$$

By assumption A has rank n so that n of the a_j are linearly independent. Because (A.1) shows that $A^{\mathrm{T}}A$ is independent of how the rows of A are ordered, we can assume with no loss of generality that the first n rows are linearly independent. We will prove that $A^{\mathrm{T}}A$ is nonsingular by showing that the only solution of $A^{\mathrm{T}}Ax = 0$ is $x = 0$.

Write

$$A = \begin{bmatrix} A_1 \\ A_2 \end{bmatrix} \begin{matrix} \}n \\ \}m - n \end{matrix}$$

where A_1 is composed of the first n linearly independent a_j^T. Then A_1 is nonsingular. Now, if x is a solution of $A^TAx = 0$, then

$$0 = x^TA^TAx = \|Ax\|^2 = \|A_1x\|^2 + \|A_2x\|^2$$

and this implies $A_1x = 0$, and since A_1 is nonsingular, $x = 0$.

APPENDIX II.B A Matrix Inversion Lemma

Lemma

$$[\Lambda_1 - \Lambda_{12}\Lambda_2^{-1}\Lambda_{21}]^{-1} = \Lambda_1^{-1} + \Lambda_1^{-1}\Lambda_{12}[\Lambda_2 - \Lambda_{21}\Lambda_1^{-1}\Lambda_{12}]^{-1}\Lambda_{21}\Lambda_1^{-1},$$

where it is assumed that the matrices are consistently dimensioned and that the required matrix inverses exist.

Proof: Consider

$$\begin{bmatrix} \Lambda_1 & \Lambda_{12} \\ \Lambda_{21} & \Lambda_2 \end{bmatrix}^{-1} = \begin{bmatrix} P_1 & P_{12} \\ P_{21} & P_2 \end{bmatrix}$$

Expansion of $\Lambda P = I$ results in

$$\Lambda_1 P_1 + \Lambda_{12}P_{21} = I, \qquad \Lambda_1 P_{12} + \Lambda_{12}P_2 = 0$$

$$\Lambda_{21}P_1 + \Lambda_2 P_{21} = 0, \qquad \Lambda_{21}P_{12} + \Lambda_2 P_2 = I$$

Since Λ_1 and Λ_2 are assumed nonsingular, we can write

$$P_{21} = -\Lambda_2^{-1}\Lambda_{21}P_1, \qquad P_{12} = -\Lambda_1^{-1}\Lambda_{12}P_2 \tag{a}$$

$$P_1 = [\Lambda_1 - \Lambda_{12}\Lambda_2^{-1}\Lambda_{21}]^{-1}, \qquad P_2 = [\Lambda_2 - \Lambda_{21}\Lambda_1^{-1}\Lambda_{12}]^{-1} \tag{b}$$

Now operate on the right side and expand $P\Lambda = I$ to obtain

$$P_1\Lambda_1 + P_{12}\Lambda_{21} = I \Rightarrow P_1 = \Lambda_1^{-1} - P_{12}\Lambda_{21}\Lambda_1^{-1} \tag{c}$$

Equations (a)–(c) together give the desired result.

The general identity of this matrix lemma is useful in relating the least squares (normal equation) solution with the Kalman filter (cf. Ho (1963)).

Further, results (a) and (b) are useful in themselves. In the application of information filtering, one frequently is interested in the covariance of only a subset of the parameters of the problem. Equations (a) and (b) show how such covariances can be computed. For example, P_1 could be the covariance of the position and velocity components of a spacecraft, and P_2 could be the covariance of various other parameters which affect the spacecraft state.

Remark: Householder [3] attributes this matrix inversion identity to Woodbury (1950). It is Ho (1963), however, who introduced it to the estimation community. Because of its utility, this relation or some derivative of it is frequently rediscovered, and it is suspected that the result goes back even further than Woodbury.

APPENDIX II.C Data Processing Using the Information Matrix

In this appendix we indicate how, using the information matrix, one can recursively addend data, i.e., we show how the array $(\Lambda_j\ d_j)$, with $d_j = \Lambda_j x_j$, is to be combined with the data $(A_{j+1}\ z_{j+1})$, where $z_{j+1} = A_{j+1}x + \nu_{j+1}$.

Initialization

$$\Lambda_0, \quad d_0 = \Lambda_0 x_0; \qquad (\Lambda_0 \quad d_0)$$

Observation Processing $(z_j = A_j x + \nu_j)$

$$\left.\begin{aligned} \Lambda_j &= \Lambda_{j-1} + A_j^{\mathrm{T}} A_j \\ d_j &= d_{j-1} + A_j^{\mathrm{T}} z_j \end{aligned}\right\} \quad (\Lambda_j \quad d_j) = (\Lambda_{j-1} \quad d_{j-1}) + A_j^{\mathrm{T}}(A_j \quad z_j)$$

When required, the estimate and covariance are obtained from the array $(\Lambda\ d)$ via

$$x_{\mathrm{est}} = \Lambda^{-1} d, \qquad \mathrm{cov}(x_{\mathrm{est}}) = \Lambda^{-1}$$

The recursive information matrix processing algorithm given here is computationally efficient and requires relatively little storage space; essentially only Λ (symmetric) and d are required. It is ideally suited to problems that involve large numbers of observations and where estimates and/or covariances are needed relatively infrequently. Problems in which numeric accuracy is critical require more sophisticated algorithms (cf. Chapters IV and V and the references therein).

APPENDIX II.D Data Processing Using the Kalman Algorithm

The Kalman data processing algorithm is designed to modify an estimate covariance pair to include the effect of an additional scalar observation. The recursive structure of the algorithm is indicated below and is then FORTRAN mechanized.

Initialization

x_0, P_0 (initial estimate and estimate error covariance, respectively)

Observation Processing† $(z_j = A_j x + \nu_j)$

$$v_j = P_j A_j^{\mathrm{T}}$$

$$\sigma_j = A_j v_j + 1 \qquad \text{(predicted residual variance)}$$

$$K_j = v_j / \sigma_j \qquad \text{(Kalman gain)}$$

$$\tilde{\nu}_j = z_j - A_j x_j \qquad \text{(predicted residual)}$$

$$x_{j+1} = x_j + K_j \tilde{\nu}_j \qquad \text{(state update)}$$

$$\overline{P}_{j+1} = P_j - K_j v_j^{\mathrm{T}} \qquad \text{(optimal covariance update)}$$

$$\bar{v}_j = \overline{P}_{j+1} A_j^{\mathrm{T}}$$

$$P_{j+1} = \left(\overline{P}_{j+1} - \bar{v}_j K_j^{\mathrm{T}}\right) + K_j K_j^{\mathrm{T}} \qquad \text{(stabilized covariance update)}^{\ddagger}$$

where x_j and P_j are the estimate and estimate error covariance after processing j observations.

Input x, P *A priori* estimate and covariance
z, A Measurement and measurement coefficients

†Scalar observation processing.
‡This equation arrangement enhances numerical stability.

Output x, P Updated estimate and covariance
K Kalman gain vector

Remark: The symbol @ precedes internal FORTRAN comments.

```
   σ=1.                                 @ σ=E(ν²)=1.
   δ=z
   DO 10 I=1, N
   v(I)=0.
   DO 5 J=1, N
 5 v(I)=v(I)+P(I, J) * A(J)
   δ=δ-A(I) * x(I)
10 σ=σ+A(I) * v(I)
```

Comment: $v = PA^{\mathrm{T}}$, $\sigma = APA^{\mathrm{T}} + 1$, and $\delta = z - Ax$ have been computed.

```
   σ = 1./σ
   DO 20 I = 1, N
   K(I) = v(I) * σ                      @ Kalman gain
   x(I) = x(I) + K(I) * δ               @ state update
   DO 20 J = I, N
   P(I, J) = P(I, J) - K(I) * v(J)      @ conventional update
20 P(J, I) = P(I, J)                    @ utilize symmetry
```

Comment: Kalman gain, updated estimate and optimal covariance have been computed.

```
   DO 30 I = 1, N
   v(I) = 0.
   DO 30 J = 1, N
30 v(I) = v(I) + P(I, J) * A(J)
   DO 40 J = 1, N
   DO 40 I = 1, J
   s =.5 * (P(I, J) - v(I) * K(J) + P(I, J) - v(J) * K(I))
   P(I, J) = s + K(I) * K(J),           @ stabilized update
40 P(J, I) = P(I, J)                    @ utilize symmetry
```

Comment: P is the updated covariance.

Remarks: (1) Observe how the variables are redundantly used. Writing over old variables economizes on storage and results in a more compact implementation.

(2) Note that our FORTRAN implementation is purposely not strict. We freely use Greek letters and Latin lowercase letters because FORTRAN multisymbol names obfuscate the code and lowercase letters are

often associated with vector terminology. The reader who is familiar with FORTRAN will have no difficulty implementing our codes.

(3) This mechanization has proven to be numerically better conditioned than the usual Joseph stabilized formula.

APPENDIX II.E Data Processing Using the Potter Algorithm

In this appendix we order the equations that are necessary for implementation of Potter's data processing method. A FORTRAN mechanization is included for the convenience of the reader interested in implementing this algorithm.

Initialization

x_0, S_0 (initial estimate and square root error covariance,† respectively)

Observation Processing‡ $(z_j = A_j x + \nu_j)$

$$v_j^{\mathrm{T}} = A_j S_j$$

$$\bar{\sigma}_j = 1./(v_j^{\mathrm{T}} v_j + 1.) \qquad \text{(predicted residual variance inverse)}$$

$$\bar{K}_j = S_j v_j \qquad \text{(unweighted Kalman gain)}$$

$$\delta_j = z_j - A_j x_j \qquad \text{(predicted residuals)}$$

$$x_{j+1} = x_j + \bar{K}_j(\delta_j \bar{\sigma}_j) \qquad \text{(updated state estimate)}$$

$$\gamma_j = \bar{\sigma}_j/(1.+\sqrt{\bar{\sigma}_j}\,)$$

$$S_{j+1} = S_j - (\gamma_j \bar{K}_j) v_j^{\mathrm{T}} \qquad \text{(square root covariance update)}$$

Input x, S *A priori* estimate and square root covariance
z, A Measurement and measurement coefficients

†$P_0 = S_0 S_0^{\mathrm{T}}$. If P_0 is diagonal, then one can take $S_0 = \mathrm{Diag}((P(1,1))^{1.2}, \ldots, (P(n,n))^{1.2})$ otherwise one can use the Cholesky algorithm (cf. Appendix III.B) to compute S_0.

‡Scalar observation processing.

Output x, S Updated estimate and square root covariance

```
      σ = 1.                               @ σ = E(ν²) = 1.
      δ = z
      DO 10 I = 1, N
      v(I) = 0.
      DO 5 J = 1, N
 5    v(I) = v(I) + S(J, I) * A(J)
      δ = δ − A(I) * x(I)
10    σ = σ + v(I) ** 2
```

Comment: $v = S^T A^T$, $\sigma = 1 + v^T v$, and $\delta = z - Ax$ have been computed.

```
      σ = 1./σ                             @ σ̄, reciprocal of the innovations
                                             variance
      δ = δ * σ                            @ scaled computed residual
      γ = σ/(1.+SQRT(σ))
      DO 20 I = 1, N
      α = 0.
      DO 15 J = 1, N
15    α = α + S(I, J) * v(J)               @ α = K̄(I), the unweighted gain
      x(I) = x(I) + α * δ                  @ updated estimate
      α = α * γ
      DO 20 J = 1, N
20    S(I, J) = S(I, J) − α * v(J)         @ updated square root covariance
```

References

[1] Deutsch, R., "Estimation Theory." Prentice-Hall, Englewood Cliffs, New Jersey (1965).

This text, one of the first books on this subject, contains excellent historical background. Chapters 4, 5, and 8 contain supplementary material on least squares.

[2] Battin, R. H., "Astronautical Guidance," pp. 338–339. McGraw-Hill, New York (1964).

[3] Householder, A. S., "The Theory of Matrices in Numerical Analysis." Ginn (Blaisdell), Waltham, Massachusetts (1964).

This classic text on numerical linear algebra is tersely written but contains much to recommend it to the computationally oriented estimation practitioner.

[4] Carlson, N. A., Fast triangular factorization of the square root filter, *AIAA J.* **11**, No. 9, 1259–1265 (1973).

The triangular factorization reported here is clever and promising. Unfortunately the paper contains several exaggerated and unsubstantiated claims, and these cast aspersions on the paper's overall credibility.

[5] Gill, P. E., Golub, G. H., Murray, W., and Saunders, M. A., Methods for modifying matrix factorizations, *Math. Comp.* **28**, No. 126, 505–535 (1974).

This paper surveys some of the lesser known methods of updating square root matrices. In particular it contains an elegant, practical method for updating Cholesky factors of $LDL^{\mathrm{T}} + aa^{\mathrm{T}}$, with L unit lower triangular, D diagonal, and a a vector. It also gives a stable and efficient construction based on Householder orthogonal transformations.

[6] Athans, M., and Schweppe, F. C., Gradient matrices and matrix calculations. MIT Lincoln Lab. Tech. Note 1965-53 (November 1965). Available from the Clearinghouse for Federal Scientific and Technical Information, No. AD 624426.

[7] Athans, M., The matrix minimum principle. *Information and Control* **11**, 592–606 (1968).

III

Positive Definite Matrices, the Cholesky Decomposition, and Some Applications

In this chapter attention is focused on the factorization of a positive definite (PD) matrix. Nonsingular covariance matrices are PD and for this reason such matrices are important for estimation analyses. In this chapter we review some properties of PD matrices that are useful for estimation applications, derive useful PD matrix triangular factorizations, and lastly apply PD factorization to several estimation problems. Appendices are included with FORTRAN mechanizations of useful PD factorization algorithms.

III.1 Positive Definite Matrices

A symmetric matrix P, $P(n, n)$,† is *positive definite* (PD) iff $x^{\mathrm{T}}Px > 0$ for all n vectors x with $\|x\| > 0$.

The expression $x^{\mathrm{T}}Px$ is a *quadratic form* in the components of x. It is not unreasonable for us to define PD matrices in terms of their associated quadratic forms because our principal contact with such matrices is through our study of least squares problems, and these are quadratic forms. An example of a quadratic form for $n = 3$ is

$$\begin{aligned} x^{\mathrm{T}}Px &= \sum_{i,j=1}^{n} P(i,j)x_i x_j \\ &= P(1,1)x_1^2 + 2P(1,2)x_1x_2 + 2P(1,3)x_1x_3 \\ &\qquad + P(2,2)x_2^2 + 2P(2,3)x_2x_3 \\ &\qquad + P(3,3)x_3^2 \end{aligned}$$

†That is, P has dimension n.

Observe how one can write the quadratic form directly from the matrix representation of P.

III.2 Properties of PD Matrices

In this section we briefly summarize some of the properties of PD matrices that are useful for estimation analyses. When P is a symmetric matrix the notation $P > 0$ will mean that P is PD.

Property A $P > 0$ iff all the eigenvalues of P are positive.

Property B If $P > 0$, the diagonal elements of P are positive.

Property C If matrix M is nonsingular and $P > 0$, then $M^{\mathrm{T}}PM$ is also PD.

Property D If $P > 0$, then P^{-1} exists and $P^{-1} > 0$.

Property E If $P > 0$, the matrix obtained by deleting a row and corresponding column from P is also PD.

Property F If $P > 0$, and $\rho(i, j) = P(i, j)/[P(i, i)P(j, j)]^{1/2}$, $i, j = 1, \ldots, n$, with $i \neq j$; then $|\rho(i, j)| < 1$. $\rho(i, j)$ are called *correlation coefficients*.

Proof of A: Suppose that (λ, u) is an eigenvalue-eigenvector pair corresponding to P and $P > 0$. Then

$$\begin{aligned} 0 &< u^{\mathrm{T}}Pu && \text{(definition of PD)} \\ &= u^{\mathrm{T}}(Pu) \\ &= u^{\mathrm{T}}(\lambda u) && (u \text{ is an eigenvector}) \\ &= \lambda u^{\mathrm{T}}u = \lambda \|u\|^2 \end{aligned}$$

and this shows that $\lambda > 0$.

To show the converse property, that if all the eigenvalues of P are positive, then $P > 0$, we consider

$$J = \min_{x \neq 0} \left(\frac{x^{\mathrm{T}}Px}{x^{\mathrm{T}}x} \right) = \min_{x \neq 0} \left(\frac{x^{\mathrm{T}}}{\|x\|} P \frac{x}{\|x\|} \right) = \min_{\|\hat{x}\| = 1} (\hat{x}^{\mathrm{T}}P\hat{x}) \tag{2.1}$$

To compute J, one introduces a Lagrange multiplier μ, and writes

$$\bar{J} = x^{\mathrm{T}}Px - \mu(x^{\mathrm{T}}x - 1)$$

Then the first variation $\delta \bar{J}$ is

$$\delta \bar{J} = x^{\mathrm{T}} P\, \delta x + (\delta x)^{\mathrm{T}} P x - \mu x^{\mathrm{T}}\, \delta x - \mu(\delta x)^{\mathrm{T}} x - \delta\mu(x^{\mathrm{T}} x - 1)$$
$$= 2(\delta x)^{\mathrm{T}}(Px - \mu x) - \delta\mu(x^{\mathrm{T}} x - 1)$$

Thus the first variation of the quadratic form $\bar{J}$ vanishes when

$$Px = \mu x \qquad \text{and} \qquad x^{\mathrm{T}} x = 1 \tag{2.2}$$

From this we conclude that μ is an eigenvalue of P, and the minimum is achieved at an eigenvector x_μ. Therefore,

$$J = \min_{\|x\|=1} x^{\mathrm{T}} P x$$

$= \min(x_\mu^{\mathrm{T}} P x_\mu)$ (x_μ is an eigenvector of P and $\|x_\mu\| = 1$)

(We can take the minimization over this subset of x values because the minimum is achieved at $x = x_\mu$ for some eigenvector x_μ.)

$= \min_{\|x_\mu\|=1} (x_\mu^{\mathrm{T}} \mu x_\mu)$ (μ is the eigenvalue corresponding to x_μ)

$= \min_{\|x_\mu\|=1} (\mu)$

$= \min(\mu)$ (μ an eigenvalue of P)

$= \mu_{\min} > 0$ (by hypothesis)

Since it has been shown that

$$\min_{x \neq 0} x^{\mathrm{T}} P x / \|x\|^2 > 0$$

it follows that

$$x^{\mathrm{T}} P x > 0 \qquad \text{for all } x \neq 0$$

i.e., $P > 0$.

Proof of B: Let e_j be the jth coordinate vector, with 1 in the jth coordinate and zeros elsewhere. Then

$$0 < e_j^{\mathrm{T}} P e_j \qquad \text{(definition of PD)}$$
$$= P(j, j)$$

and this proves Property B.

Proof of C: Let $x = My$. Since M is nonsingular there is a one-to-one correspondence between the x's and the y's. Now, observe that

$$x^{\mathrm{T}} P x = y^{\mathrm{T}} M^{\mathrm{T}} P M y$$

so that as x ranges over all values different from zero, y will range over all values different from zero; and this shows that $M^{\mathrm{T}}PM > 0$.

Proof of D: The only solution to $Px = 0$ is the trivial solution $x = 0$ because if x were a nontrivial solution, then $x^{\mathrm{T}}(Px) = x^{\mathrm{T}} \cdot 0 = 0$, and this would contradict the PD property. Thus P is nonsingular. To see that P^{-1} is symmetric when P is, note that

$$PP^{-1} = I \Rightarrow (PP^{-1})^{\mathrm{T}} = (P^{-1})^{\mathrm{T}}P^{\mathrm{T}} = (P^{-1})^{\mathrm{T}}P = I$$

and

$$(P^{-1})^{\mathrm{T}}P = I \Rightarrow (P^{-1})^{\mathrm{T}} = P^{-1}$$

Remark: By an abuse of notation we will in the sequel often write $P^{-\mathrm{T}}$ instead of $(P^{-1})^{\mathrm{T}}$, where P is an arbitrary nonsingular matrix. Since $(P^{-1})^{\mathrm{T}} = (P^{\mathrm{T}})^{-1}$, there is no danger of ambiguity or confusion.

That P^{-1} is PD follows from Property C, with $M = P^{-1}$, because

$$M^{\mathrm{T}}PM = P^{-1}PP^{-1} = P^{-1}$$

Proof of E: Suppose that column j and row j are deleted from P. We call this reduced matrix $\overline{P}$. Evidently,

$$x^{\mathrm{T}}Px = \overline{x}^{\mathrm{T}}\overline{P}\,\overline{x}$$

for those vectors x whose jth component is zero, and where $\overline{x}$ is the reduced x, with the jth component deleted. This equation proves that $\overline{P}$ is PD.

Remark: Property E implies that the matrix resulting from deletion of any number of rows and corresponding columns of a PD matrix is PD.

Proof of F: If one sets all the components of x to zero except $x_i = 1$ and $x_j = \lambda$, then a direct expansion of the quadratic form $x^{\mathrm{T}}Px$ gives

$$0 < x^{\mathrm{T}}Px = \left[(P(j,j)\lambda + P(i,j))^2 + P(i,i)P(j,j) - P(i,j)^2\right]/P(j,j)$$

Since this expression is positive for all values of λ, and $P(j,j)$ is positive, it follows that

$$P(i,i)P(j,j) - P(i,j)^2 > 0$$

and this proves Property F.

Note that

$$\det\begin{pmatrix} P(i,i) & P(i,j) \\ P(i,j) & P(j,j) \end{pmatrix} = P(i,i)P(j,j) - P(i,j)^2$$

i.e., the determinant of every two-dimensional PD matrix is positive. Although it is not relevant to our applications we note that one can prove inductively that the determinant of every "k-dimensional" PD matrix is positive.

An Aside: It is sometimes suggested, evidently motivated by Property F, that a method to force positivity of the computed covariance, Eq. (II.6.13), is to adjust the off-diagonal terms so that correlations are bounded away from unity. To see that this is not a very sure-fire method of ensuring positivity consider

$$P = \begin{bmatrix} 1 & 0.5 - \delta & 0.5 - \delta \\ 0.5 - \delta & 1 & -0.5 + \delta \\ 0.5 - \delta & -0.5 + \delta & 1 \end{bmatrix}$$

Requiring that the correlations be bounded away from unity implies only that $-0.5 < \delta < 1.5$. However, $\det(P) = 2\delta(1.5 - \delta)^2$, and thus in addition one must require that $\delta > 0$. Evidently requiring that the correlations be less than unity is not sufficient. ■

III.3 Matrix Square Roots and the Cholesky Decomposition Algorithm

If matrix P can be written as

$$P = SS^{\mathrm{T}}$$

with S a square matrix, we say that S is a *square root* of P.

Remark: This definition, although frequently used, is not universal. Some authors generalize and allow S to be rectangular, while others restrict S to be symmetric, and still others allow S to have complex entries. In the latter case one would say that S is a (complex) square root of P if $P = SS^H$, where S^H is the complex conjugate transposed matrix of S.

Square root matrices, when they exist, are nonunique because if S is one square root of P and T is orthogonal ($TT^{\mathrm{T}} = I$), then ST is also a square root of P. (Orthogonal matrices will be discussed in more detail in Chapter IV.) We now set about showing that every PD matrix has a square root. Our proof, based on the elementary process of completing the square, is constructive. In the interest of making the construction in the general case more transparent, we first examine the case of $n = 3$:

$$\begin{aligned} x^{\mathrm{T}}Px = {} & P(1,1)x_1^2 + 2(P(1,2)x_2 + P(1,3)x_3)x_1 \\ & + P(2,2)x_2^2 + 2P(2,3)x_2x_3 \\ & + P(3,3)x_3^2 \end{aligned} \tag{3.1}$$

$$\begin{aligned} = {} & \left[P(1,1)^{1/2}x_1 + (P(1,2)x_2 + P(1,3)x_3)/P(1,1)^{1/2}\right]^2 \\ & - \left[(P(1,2)x_2 + P(1,3)x_3)/P(1,1)^{1/2}\right]^2 \\ & + P(2,2)x_2^2 + 2P(2,3)x_2x_3 \\ & + P(3,3)x_3^2 \end{aligned} \tag{3.2}$$

Note that we have completed the square of the x_1 quadratic term, and that by Property B, $P(1, 1)$ is positive. The expressions in (3.2) suggest that additional notation might help, i.e., we set†

$$\begin{aligned} L(1,1) &= P(1,1)^{1/2} \\ L(2,1) &= P(2,1)/L(1,1) \\ L(3,1) &= P(3,1)/L(1,1) \end{aligned} \tag{3.3}$$

and let

$$y_1 = \sum_{i=1}^{3} L(i,1)x_i \tag{3.4}$$

In terms of this notation we can write

$$\begin{aligned} x^{\mathrm{T}}Px &= y_1^2 + (P(2,2) - L(2,1)L(2,1))x_2^2 \\ &\quad + 2(P(2,3) - L(2,1)L(3,1))x_2x_3 \\ &\quad + (P(3,3) - L(3,1)L(3,1))x_3^2 \\ &= y_1^2 + \bar{P}(2,2)x_2^2 + 2\bar{P}(2,3)x_2x_3 + \bar{P}(3,3)x_3^2 \end{aligned} \tag{3.5}$$

where†

$$\bar{P}(i,j) = P(i,j) - L(i,1)L(j,1) \qquad \begin{cases} i = 2, 3 \\ j = i, 3 \end{cases} \tag{3.6}$$

†Recall that $P(j, 1) = P(1, j)$. The equations are written in terms of elements situated below the diagonal because this appears consistent with our intent here of constructing a lower triangular factorization and because it is nice (for purposes of computer implementation) to involve only the lower triangular elements of P in the construction.

We now complete the square on the x_2 variables, i.e.,

$$\begin{aligned}\bar{P}(2,2)x_2^2 + 2\bar{P}(2,3)x_2x_3 + \bar{P}(3,3)x_3^2 \\ = \left[\bar{P}(2,2)^{1/2}x_2 + \left(\bar{P}(2,3)/\bar{P}(2,2)^{1/2}\right)x_3\right]^2 \\ - \left[\left(\bar{P}(2,3)/\bar{P}(2,2)^{1/2}\right)x_3\right]^2 + \bar{P}(3,3)x_3^2\end{aligned} \tag{3.7}$$

Note that $\bar{P}(2, 2)$ is positive. To see this, choose $x_2 = 1$, $x_3 = 0$ and choose x_1 so that $y_1 = 0$. Then for this x, $x^{\mathrm{T}}Px = \bar{P}(2, 2)$.

Consistent with (3.3) and (3.4), we set

$$L(2,2) = \bar{P}(2,2)^{1/2}, \qquad L(3,2) = \bar{P}(3,2)/L(2,2) \tag{3.8}$$

and

$$y_2 = L(2,2)x_2 + L(3,2)x_3 \tag{3.9}$$

If we now set†

$$L(3,3) = \left(\bar{P}(3,3) - L(3,2)L(3,2)\right)^{1/2}$$

and

$$y_3 = L(3,3)x_3$$

then we have

$$x^{\mathrm{T}}Px = \sum_{i=1}^{3} y_i^2 = y^{\mathrm{T}}y \tag{3.10}$$

However,

$$y = \begin{bmatrix} L(1,1) & L(2,1) & L(3,1) \\ 0 & L(2,2) & L(3,2) \\ 0 & 0 & L(3,3) \end{bmatrix} x \tag{3.11}$$

so that

$$y^{\mathrm{T}}y = x^{\mathrm{T}}LL^{\mathrm{T}}x \tag{3.12}$$

where the labeling of the matrix entries has been arranged so that $y = L^{\mathrm{T}}x$, L being a lower triangular matrix. Since (3.12) is also equal to $x^{\mathrm{T}}Px$, we have $P = LL^{\mathrm{T}}$, i.e., we have constructed a lower triangular square root of P.

One can generalize from the example and conclude that if $P = LL^{\mathrm{T}}$, then the change of variable $y = L^{\mathrm{T}}x$ converts the quadratic form $x^{\mathrm{T}}Px$ to a sum of squares. Triangular factorization in the general case follows the case for $n = 3$.

†Using reasoning analogous to that used to show that $P(2, 2) > 0$, one can show that $\bar{P}(3, 3) - L(3, 2)L(3, 2)$ is positive, i.e., choose $x_3 = 1$, and x_1 and x_2 such that y_1 and y_2 are zero. Then for this x, $x^{\mathrm{T}}Px = \bar{P}(3, 3) - L(3, 2)^2$.

Theorem (Lower Triangular Cholesky Decomposition) If $P > 0$ it has a lower triangular factorization, $P = LL^{\mathrm{T}}$. The factorization with positive diagonal elements is given by the following algorithm:

For $j = 1, \ldots, n-1$, recursively cycle through the following ordered set of equations:

$$\left.\begin{aligned} L(j,j) &= P(j,j)^{1/2} \\ L(k,j) &= P(k,j)/L(j,j), \qquad k = j+1, \ldots, n \\ P(i,k) &:= P(i,k) - L(i,j)L(k,j) \quad \begin{cases} k = j+1, \ldots, n \\ i = k, \ldots, n \end{cases} \end{aligned}\right\} \tag{3.13}$$

and then $L(n, n) = P(n, n)$, where $n = \dim P$.

Remark: The symbol ":=" may be read "is replaced by" and means that the quantity on the left is to be replaced by the result on the right. In general, it is poor mathematical style to redefine symbols in such a fashion. Once, however, the computation of the modified P terms has been completed, there will be no further need to address the original P terms. Moreover the new P terms will be associated with a modified quadratic form in the same way as in the original problem. Our intent, therefore, is to encourage the reader to "think sequentially" and develop the habit of reusing name tags that retain their significance. The value of this technique is that unwieldy notation is avoided and at the same time the results are in a form that translates directly into a computer mechanization (cf. Appendices III.A–III.C).

Proof:† The proof duplicates the $n = 3$ case. To begin we write

$$\begin{aligned} x^{\mathrm{T}}Px &= P(1,1)x_1^2 + 2\left(\sum_{j=2}^{n} P(1,j)x_j\right)x_1 \\ &\quad + \sum_{i,j=2}^{n} P(i,j)x_i x_j \qquad \text{(cf. Eq. (3.1))} \\ &= \left[P(1,1)^{1/2}x_1 + \sum_{j=2}^{n} P(1,j)/P(1,1)^{1/2}x_j\right]^2 \\ &\quad - \left[\sum_{j=2}^{n}\left(P(1,j)/P(1,1)^{1/2}\right)x_j\right]^2 \\ &\quad + \sum_{i,j=2}^{n} P(i,j)x_i x_j \qquad \text{(cf. Eq. (3.2))} \end{aligned}$$

†This proof, based on the geometric significance of the triangular factorization, is given as an exercise by Noble [2].

and we set

$$L(1, 1) = P(1, 1)^{1/2}$$

$$L(j, 1) = P(j, 1)/L(1, 1), \qquad j = 2, \ldots, n \quad \text{(cf. Eq. (3.3))}$$

$$y_1 = \sum_{j=1}^{n} L(j, 1)x_j \qquad \text{(cf. Eq. (3.4))}$$

Rewriting $x^T P x$ in terms of y_1 and the L's gives

$$x^T P x = y_1^2 + \sum_{i,j=2}^{n} [P(i, j) - L(i, 1)L(j, 1)]x_i x_j \tag{3.14}$$

To keep the notation from getting unwieldy and because we prefer notation related to computer implementation, we write the new values of $P(i, j)$ over the old ones, i.e.,

$$P(i, j) := P(i, j) - L(i, 1)L(j, 1), \qquad i, j = 2, \ldots, n$$

Now we have the same problem with which we began, except that the quadratic form now only involves $x_2, \ldots, x_n$, i.e.,

$$\sum_{i,k=2}^{n} P(i, k)x_i x_k$$

One shows that the new $P(2, 2)$ is positive by noting that $x^T P x = P(2, 2)$ if x is given by $x_2 = 1$, $x_j = 0$, $j > 2$, and x_1 is such that $y_1 = 0$.

Completing the square on x_2 gives $L(2, i)$, $i = 2, \ldots, n$ and y_2. This procedure is repeated until all of the x's have been replaced by y's with

$$y = \begin{bmatrix} L(1,1) & & \cdots & L(n,1) \\ & L(2,2) & & \vdots \\ & & \ddots & \\ 0 & & & L(n,n) \end{bmatrix} x$$

and

$$x^T P x = y^T y = x^T \begin{bmatrix} L(1,1) & & & 0 \\ \vdots & L(2,2) & & \\ & & \ddots & \\ L(n,1) & & \cdots & L(n,n) \end{bmatrix} \begin{bmatrix} & & & \\ & & & \\ & & & \\ & & & \end{bmatrix}^T x$$

This completes the proof.

Remarks: (1) Once it has been proven that P has a lower triangular square root, one could determine it a column at a time by simply comparing elements of LL^{T} with corresponding elements of P. In fact, most computer mechanizations follow this procedure.

(2) It is not difficult to show that if L_1 and L_2 are two lower triangular factorizations of P, then

$$L_1 = L_2 \,\mathrm{Diag}(\pm 1, \ldots, \pm 1)$$

i.e., one could take $L(j,j) = \pm P(j,j)^{1/2}$. We have chosen the unique square root corresponding to positive diagonal L elements.

A perusal of the Cholesky algorithm reveals that it involves n scalar square roots and they appear on the diagonals and divide the columns. This suggests factoring L and writing $P = \overline{L}\,D\,\overline{L}^{\mathrm{T}}$ with $\overline{L}$ unit lower triangular and D diagonal. Such a representation is free of square roots and has application in matrix inversion and parameter estimation (cf. Chapter V).

Corollary (Square Root Free Cholesky Decomposition) If $P > 0$, then $P = \overline{L}\,D\,\overline{L}^{\mathrm{T}}$, where $\overline{L}$ is lower triangular with unit diagonal elements and $D = \mathrm{Diag}(d_1, \ldots, d_n)$. The elements of $\overline{L}$ and D are given by the following algorithm:

For $j = 1, \ldots, n-1$, recursively cycle through the following ordered set of equations:

$$d_j = P(j,j), \qquad \overline{L}(j,j) = 1 \tag{3.15}$$

$$P(i,k) := P(i,k) - \overline{L}(i,j)P(k,j) \qquad \begin{cases} k = j+1, \ldots, n \\ i = k, \ldots, n \end{cases} \tag{3.16}$$

$$\overline{L}(k,j) = P(k,j)/d_j, \qquad k = j+1, \ldots, n \tag{3.17}$$

and then $d_n = P(n,n)$. The result can be obtained by mimicking the Cholesky theorem derivation, or more directly, by identifying $d_j = L(j,j)^2$ and $\overline{L}(i,j) = L(i,j)/L(j,j)$ and manipulating result (3.13).

Upper triangular factorization algorithms can be obtained by mimicking the developments of this section. The corresponding upper triangular Cholesky factorizations are given in Appendix III.A. Upper and lower triangular factorizations require the same amount of computation so that deciding which factorization to use depends on the demands of the application.

Our investigations and algorithms will involve factors of the error covariance matrix, with $P = SS^{\mathrm{T}}$, and factors of the information matrix P^{-1}, written as $R^{\mathrm{T}}R$. The computation of R will often put it in an upper triangular form and since $R = S^{-1}$ this motivates us to write S in upper triangular form (because matrix inversion preserves triangularity). The reason that our developments in this section have favored the lower triangular factorization is that this representation is useful for certain types of estimation problems. Our preference for upper triangular factorizations will be expressed in the development of Section III.4 and in the representative FORTRAN mechanizations given in Appendices III.A and III.C.

Remark: Numerical analysts have studied the computational stability and accuracy of PD matrix triangular factorizations. These factorizations are reasonably stable and accurate and are generally unaffected by pivot element strategies, which, if necessary, would complicate implementation (cf. Wilkinson [3], and also Wilkinson (1965a, pp. 231–233)). Despite this positive endorsement it happens that numerical difficulties are not uncommon. In most cases the problem can be traced to roundoff errors involved with the formation of a nearly singular PD matrix.

Matrices that are symmetric or triangular can be described by $n(n + 1)/2$ of their elements. Analysts dealing with problems involving many variables can save storage by storing the matrix in the form of a vector, i.e.,

$$\begin{bmatrix} \overset{\textcircled{1}}{P(1,1)} & \overset{\textcircled{2}}{P(1,2)} & \overset{\textcircled{4}}{P(1,3)} & \\ & \overset{\textcircled{3}}{P(2,2)} & \overset{\textcircled{5}}{P(2,3)}, & \text{etc}. \\ & & \overset{\textcircled{6}}{P(3,3)} & \end{bmatrix}$$

This columnwise labeling of the upper triangular portion of the matrix is called *vector subscripting*, and is well suited to problems involving the augmentation of additional matrix columns. It is not difficult to relate the double dimension subscript with the corresponding vector subscript, i.e., $(i, j) \rightleftarrows (j(j - 1)/2 + i)$. Since matrix multiplication involves successive rows and columns, one can often arrange an efficient algorithm implementation so that the vector-subscripted indices are computed recursively [i.e., $(i, j + 1) \rightleftarrows ((i, j) + j)$, column incrementing and $(i + 1, j) \rightleftarrows ((i, j) + 1)$, row incrementing]. Appendices III.A and IV.B contain examples illustrating how vector subscripting can be incorporated into FORTRAN mechanizations.

III.4 Rank One Modification of the Cholesky Factorization[†]

Many of our intended filtering applications involve matrices of the form

$$\bar{P} = P + caa^{\mathrm{T}} \tag{4.1}$$

where P is PD, c is a scalar, and a is an n vector. The simplicity of form (4.1) suggests that one seek direct methods to calculate the Cholesky factors of $\bar{P}$ when the Cholesky factors of P are given. Obvious reasons for wanting a direct factorization are reduced computation and possibly increased numeric accuracy. Interestingly enough there are a variety of factorization methods that can be applied, and selecting an appropriate method depends on whether P and/or $\bar{P}$ are PD, singular, or near singular. In this section we consider a method of computing the updated upper triangular Cholesky factors that is due to Agee and Turner [4].[‡] This method is numerically stable for $\bar{P}$ that are PD and not near singular (cf. Gill *et al.* [5]). The Agee–Turner result will be derived as a simple extension of the reasoning used in our development of the Cholesky algorithm. We will only consider the problem where P is given as $P = UDU^{\mathrm{T}}$, with U unit upper triangular, and will devise an algorithm to update these factors. The decision to use an upper triangular factorization is prompted by the fact that it is more consistent with our use of upper triangular square root information matrices (cf. Chapter V).

Remark: Formulae to update factorizations of the form $P = SS^{\mathrm{T}}$, with S triangular, can be readily obtained from this result, or perhaps more simply, by appropriately modifying our development.

Theorem (Agee–Turner PD Factorization Update) Let

$$\bar{P} = \bar{U}\bar{D}\bar{U}^{\mathrm{T}} = UDU^{\mathrm{T}} + caa^{\mathrm{T}}$$

where c is a scalar, a is an n vector, U is unit upper triangular, $D = \mathrm{Diag}(d_1, \ldots, d_n)$, and $n = \dim P$.

If $\bar{P}$ is PD. then the factors $\bar{U}$ and $\bar{D}$ can be calculated as follows:

[†]This section is of a specialized nature and can be omitted in a first reading. This material is called upon in Section V.3 when triangular factorization of the Kalman data processing algorithm is discussed.

[‡]The Agee–Turner result is actually a special case of a more general result due to Bennett (1965).

For $j = n, n-1, \ldots, 2$ recursively evaluate ordered equations (4.2)–(4.5):

$$\bar{d}_j = d_j + c_j a_j^2 \tag{4.2}$$

$$a_k := a_k - a_j U(k, j), \qquad k = 1, \ldots, j-1 \tag{4.3}$$

$$\bar{U}(k, j) = U(k, j) + c_j a_j a_k / \bar{d}_j, \quad k = 1, \ldots, j-1 \tag{4.4}$$

$$c_{j-1} = c_j d_j / \bar{d}_j \tag{4.5}$$

and then compute

$$\bar{d}_1 = d_1 + c_1 a_1^2$$

Remarks: (1) In Eq. (4.3) the first j components of a are being redefined for use in Eq. (4.4).

(2) In Eq. (4.4) the jth column of $\bar{U}$ is being defined.

(3) In Eq. (4.5) c_{j-1} is being defined for use in the next cycle of Eqs. (4.2) and (4.4), and $c_n = c$. FORTRAN implementation (cf. Appendix III.C) does not involve subscripting of c.

Proof: The structure of the algorithm is revealed by considering the associated quadratic form $x^T\bar{P}x$:

$$x^T\bar{P}x = \sum_1^n d_j v_j^2 + c(a^T x)^2 \tag{4.6}$$

$$= \sum_1^{n-1} d_j v_j^2 + (d_n + c a_n^2) x_n^2 + 2x_n \sum_1^{n-1} (d_n U(j, n) + c a_n a_j) x_j$$

$$+ d_n \left(\sum_1^{n-1} U(j, n) x_j \right)^2 + c \left(\sum_1^{n-1} a_j x_j \right)^2 \tag{4.7}$$

where v_j are the components of $v = U^T x$. To better see the fundamental step of the derivation we introduce the $n-1$ dimensional vectors $\bar{w}$, $\bar{a}$, and $\bar{x}$,

$$\bar{w}^T = (U(1, n), \ldots, U(n-1, n)), \qquad \bar{a}^T = (a_1, \ldots, a_{n-1}),$$

$$\bar{x}^T = (x_1, \ldots, x_{n-1})$$

Then (4.7) can be written as

$$x^{\mathrm{T}}\bar{P}x = \sum_{1}^{n-1} d_j v_j^2 + \bar{d}_n\left[x_n + (d_n\bar{w} + ca_n\bar{a})^{\mathrm{T}}\bar{x}/\bar{d}_n\right]^2$$

$$+ d_n(\bar{w}^{\mathrm{T}}\bar{x})^2 + c(\bar{a}^{\mathrm{T}}\bar{x})^2 - \left[(d_n\bar{w} + ca_n\bar{a})^{\mathrm{T}}\bar{x}\right]^2/\bar{d}_n \qquad (4.8)$$

When we set

$$y_n = x_n + (1/\bar{d}_n)(d_n\bar{w} + ca_n\bar{a})^{\mathrm{T}}\bar{x}$$

and combine the other $\bar{x}$ quadratic expression, we obtain

$$x^{\mathrm{T}}\bar{P}x = \bar{d}_n y_n^2 + \sum_{1}^{n-1} d_j v_j^2$$

$$+ \bar{x}^{\mathrm{T}}\left[d_n\bar{w}\,\bar{w}^{\mathrm{T}} + c\bar{a}\,\bar{a}^{\mathrm{T}} - (d_n\bar{w} + ca_n\bar{a})(d_n\bar{w} + ca_n\bar{a})^{\mathrm{T}}/\bar{d}_n\right]\bar{x}$$

$$= \bar{d}_n y_n^2 + \left[\sum_{1}^{n-1} d_j v_j^2 + (cd_n/\bar{d}_n)\left((\bar{a} - a_n\bar{w})^{\mathrm{T}}\bar{x}\right)^2\right] \qquad (4.9)$$

Observe that the bracketed term in (4.9) is of the same form as (4.6), except that now only $n - 1$ variables are involved. Thus the inductive reduction follows recursively, and after $n - 1$ steps the quadratic form has been reduced to

$$x^{\mathrm{T}}\bar{P}x = \sum_{j=2}^{n} \bar{d}_j y_j^2 + \bar{d}_1 x_1^2$$

and since the construction has given

$$y = \bar{U}^{\mathrm{T}}x \qquad (\text{set } y_1 = x_1)$$

it follows that

$$x^{\mathrm{T}}\bar{P}x = \sum_{1}^{n} \bar{d}_j y_j^2 = x^{\mathrm{T}}\bar{U}\bar{D}\bar{U}^{\mathrm{T}}x$$

and we are done.

Gill *et al.* [5] point out that when c is negative, roundoff errors may result in some of the computed d_j terms being negative (so that the computed $\overline{P}$ will not be PD). For such problems they derive more stable algorithms, ones that are based on the theory of orthogonal transformations. The point we wish to make is that when numeric accuracy is not in jeopardy (viz., when $c > 0$) the Agee–Turner result is quicker and often of sufficient accuracy. In Chapter V, where parameter estimation algorithms are discussed, we will derive a more stable form of the Agee–Turner result that applies specifically to the Kalman covariance update Eq. (II.6.3).

We now turn attention to some applications of PD matrix factorization.

III.5 Whitening Observation Errors

Suppose that we have a set of observations

$$z = Ax + \nu \tag{5.1}$$

and that ν, the observation error, has zero mean but is correlated, i.e.,

$$E(\nu) = 0, \qquad E(\nu\nu^{\mathrm{T}}) = P_\nu$$

where P_ν is PD.

Let us write

$$P_\nu = L_\nu L_\nu^{\mathrm{T}}$$

with L_ν a lower triangular square root of P_ν. Now, if we multiply the observations (5.1) by L_ν^{-1}, we will have an observation set with unit covariance observation error, i.e.,

$$\overline{z} = \overline{A}x + \overline{\nu} \tag{5.2}$$

with

$$\overline{z} = L_\nu^{-1}z, \qquad \overline{A} = L_\nu^{-1}A, \qquad \overline{\nu} = L_\nu^{-1}\nu$$

The reason that $\overline{\nu}$ has unit covariance is explained via the following self-explanatory set of equations:

$$\begin{aligned} E(\overline{\nu}\,\overline{\nu}^{\mathrm{T}}) &= E(L_\nu^{-1}\nu\nu^{\mathrm{T}}L_\nu^{-\mathrm{T}}) \\ &= L_\nu^{-1}E(\nu\nu^{\mathrm{T}})L_\nu^{-\mathrm{T}} = L_\nu^{-1}P_\nu L_\nu^{-\mathrm{T}} \\ &= L_\nu^{-1}L_\nu L_\nu^{\mathrm{T}}L_\nu^{-\mathrm{T}} = I \end{aligned}$$

The significance of (5.2) is that it demonstrates how to construct an equivalent set of independent measurements from a set that is correlated.

An Aside: The inference associated with this point is that there is no loss of generality in assuming, in our algorithm developments, that the observations are uncorrelated and normalized. By treating this special case we avoid the complexity and inefficiency associated with the more "general" algorithm mechanizations which include provision for correlated measurements. Based on aerospace experiences (which generally involve uncorrelated observation noise) we believe that in the long run it is more efficient to use specialized algorithms and to whiten the data in those (relatively few) problems that involve correlated observations. Although we are getting a bit ahead of ourselves we mention that our general philosophy on software development is that one should whenever possible avoid burdening computer analysis programs with infrequently exercised generalities. This is especially the case when it is possible, as in the problem being discussed here, to addend a separate computation or computer program which can transform (the occasional) general problem to the requisite normalized form. ■

Let us examine two special cases of the whitening process:

(a) The observations are uncorrelated, but the ith observation error has variance σ_i^2.

In this case $L_\nu^{-1} = \mathrm{Diag}(1/\sigma_1, \ldots, 1/\sigma_m)$, and (5.2) tells us simply to scale the ith observation by $1/\sigma_i$.

Remark: Experience has shown that when data of different types are to be combined in an estimation process it is good numeric practice to introduce data normalization.

(b) The case of two correlated observations:

$$\begin{bmatrix} z_1 \\ z_2 \end{bmatrix} = \begin{bmatrix} a_1^{\mathrm{T}} \\ a_2^{\mathrm{T}} \end{bmatrix} x + \begin{bmatrix} \nu_1 \\ \nu_2 \end{bmatrix} \tag{5.3}$$

$$E(\nu\nu^{\mathrm{T}}) = \begin{bmatrix} E(\nu_1^2) & E(\nu_1\nu_2) \\ & E(\nu_2^2) \end{bmatrix} = \begin{bmatrix} \sigma_1^2 & \rho\sigma_1\sigma_2 \\ & \sigma_2^2 \end{bmatrix}^{\dagger}$$

†When a matrix is symmetric one often omits the lower triangular, redundant, terms.

where ρ is the correlation between the two observations. The lower triangular square root and its inverse are

$$L = \begin{bmatrix} \sigma_1 & 0 \\ \rho\sigma_2 & \sigma_2(1-\rho^2)^{1/2} \end{bmatrix}$$

$$L^{-1} = \begin{bmatrix} \sigma_1^{-1} & 0 \\ \dfrac{-\rho}{\sigma_1(1-\rho^2)^{1/2}} & \dfrac{1}{\sigma_2(1-\rho^2)^{1/2}} \end{bmatrix}$$

Based on these formulas, the modified observation set is obtained by scaling the first observation by its standard deviation, i.e.,

$$z_1 = a_1^{\mathrm{T}}x + \nu_1 \rightarrow \frac{1}{\sigma_1} z_1 = \frac{1}{\sigma_1} a_1^{\mathrm{T}}x + \frac{1}{\sigma_1} \nu_1 \tag{5.4a}$$

and this is of the form

$$\bar{z}_1 = \bar{a}_1^{\mathrm{T}}x + \bar{\nu}_1$$

with $\bar{\nu}_1$ normalized. Combining this with the second observation yields

$$\bar{z}_2 = \bar{a}_2^{\mathrm{T}}x + \bar{\nu}_2$$

where

$$\bar{z}_2 = \frac{\sigma_2^{-1}z_2 - \rho\bar{z}_1}{(1-\rho^2)^{1/2}} \quad \text{and} \quad \bar{a}_2^{\mathrm{T}} = \frac{\sigma_2^{-1}a_2^{\mathrm{T}} - \rho\bar{a}_1^{\mathrm{T}}}{(1-\rho^2)^{1/2}} \tag{5.4b}$$

This second observation represents the new information not contained in the first observation.

III.6 Observation Errors That Are Pairwise Correlated

Suppose that the errors in a given observation process are pairwise correlated, and for simplicity let us suppose

$$E(\nu_i^2) = \sigma^2, \qquad E(\nu_i \nu_{i+1}) = \rho_i \sigma^2 \tag{6.1}$$

In this case

$$P_\nu = \sigma^2 \begin{bmatrix} 1 & \rho_1 & & & & 0 \\ \rho_1 & 1 & \rho_2 & & & \\ & \rho_2 & 1 & \rho_3 & & \\ & & \ddots & \ddots & \ddots & \\ & & & & & \rho_{m-1} \\ 0 & & & & \rho_{m-1} & 1 \end{bmatrix}$$

and

$$L_\nu = \sigma \begin{bmatrix} \bar{\sigma}_1 & & & 0 \\ \bar{\rho}_1 & \bar{\sigma}_2 & & \\ & \ddots & \ddots & \\ 0 & & \bar{\rho}_{m-1} & \bar{\sigma}_m \end{bmatrix}$$

with $\bar{\sigma}_1 = 1$ and

$$\bar{\rho}_j = \rho_j / \bar{\sigma}_j, \qquad \bar{\sigma}_{j+1} = \left(1 - \bar{\rho}_j^2\right)^{1/2}, \qquad j = 1, \ldots, m-1 \tag{6.2}$$

For this problem we can *recursively* compute an uncorrelated sequence from the original correlated one, i.e., $\bar{z}_1 = z_1$, and for $j > 1$,

$$\bar{z}_j = (z_j - \bar{\rho}_{j-1}\bar{z}_{j-1})/\bar{\sigma}_j \tag{6.3}$$

$\{z_j\}$ are uncorrelated and have variance σ^2.

Observe that (6.3) does not require all the observations to be simultaneously available for the computation. Incidently, we note that the pairwise correlation constraint coupled with the positive definiteness of P bounds the ρ_j terms more stringently than does the correlation constraint $\rho_j^2 < 1$. For example, when all the ρ_j are equal to ρ then for $m = 3$ one must have $\rho^2 < \frac{1}{2} = 0.50$; while for $m = 4$ one must have $\rho^2 < (3 - \sqrt{5})/2 \approx 0.38$. This trend continues, however, only up to a point, and when $\rho^2 \leqslant \frac{1}{4} = 0.25$, P is positive definite for all values of m. In this case one can show that

$$\bar{\sigma}_j \geqslant \sqrt{\tfrac{1}{2} + \sqrt{\tfrac{1}{4} - \rho^2}}$$

for all $j \geqslant 1$.

III.7 Construction of Random Samples Having a Given Covariance

Suppose we wish to construct a random sample x having zero mean and covariance P. If we write

$$P = LL^{\mathrm{T}}$$

and let the components of y be n independent samples of a zero mean unit variance distribution, then

$$x = Ly$$

will have the desired statistical properties. This application is an obvious consequence of the geometrical interpretation of the matrix square root, i.e., we introduce a change of variable so that in the new coordinate system the vector components are independent, and the axes are scaled so that the components have unit variance.

Remark: Since x is a linear combination of the components of y, it will be a Gaussian random process when y is.

APPENDIX III.A Upper Triangular Matrix Factorization Algorithm

The analogue of the lower triangular factorization (3.13) is given by

Theorem (Upper Triangular Cholesky Factorization) If $P > 0$, then $P = UU^{\mathrm{T}}$, with U upper triangular. The factorization with positive diagonal elements is given by:

For $j = n, n-1, \ldots, 2$, recursively cycle through the following equations:

$$\left.\begin{aligned} &U(j,j) = P(j,j)^{1/2} \\ &U(k,j) = P(k,j)/U(j,j), \qquad k = 1, \ldots, j-1 \\ &P(i,k) := P(i,k) - U(i,j)U(k,j), \qquad \left\{\begin{aligned} &k = 1, \ldots, j-1 \\ &i = 1, \ldots, k \end{aligned}\right. \end{aligned}\right\} \tag{A.1}$$

and then set

$$U(1,1) = P(1,1)^{1/2}.$$

Proof: If P_j represents the matrix composed of the first j rows and columns of P, then

$$x^{\mathrm{T}}Px = x^{\mathrm{T}}P_n x = P(n,n)x_n^2 + 2\left(\sum_{j=1}^{n-1} P(n,j)x_j\right)x_n + (x_1, \ldots, x_{n-1})P_{n-1}\begin{pmatrix} x_1 \\ \vdots \\ x_{n-1} \end{pmatrix}$$

$$= \left[P(n,n)^{1/2}x_n + \sum_{j=1}^{n-1}\left(P(n,j)/P(n,n)^{1/2}\right)x_j\right]^2 + \sum_{i,j=1}^{n-1}\left[P(i,j) - \left(P(n,i)/P(n,n)^{1/2}\right)\left(P(n,j)/P(n,n)^{1/2}\right)\right]x_i x_j$$

$$= \left[\sum_{j=1}^{n} U(j,n)x_j\right]^2 + \sum_{i,j=1}^{n-1}\left[(P(i,j) - U(i,n)U(j,n))\right]x_i x_j$$

If we now let

$$y_n = \sum_{j=1}^{n} U(j,n)x_j$$

and

$$P(i,j) := P(i,j) - U(i,n)U(j,n), \qquad \begin{cases} j = 1, \ldots, n-1 \\ i = 1, \ldots, j \end{cases}$$

then the procedure can be repeated on P_{n-1}, etc. The algorithm follows by induction; and thus the proof is completed. ∎

We merely state the square root free upper triangular factorization, because the proof is a repetition of previous work:

$$P = UDU^{\mathrm{T}}$$

where $D = \mathrm{Diag}(d_1, \ldots, d_n)$ and U is upper triangular with unit diagonal elements.

$$\left.\begin{aligned} &\text{For } j = n, n-1, \ldots, 2 \\ &\qquad d_j = P(j,j) \\ &\quad U(j,j) = 1 \\ &\quad U(k,j) = P(k,j)/d_j, \qquad k = 1, \ldots, j-1 \\ &\quad P(i,k) := P(i,k) - U(i,j)U(k,j)d_j, \quad \begin{cases} k = 1, \ldots, j-1 \\ i = 1, \ldots, k \end{cases} \end{aligned}\right\} \quad \text{(A.2)}$$

and then

$$U(1, 1) = 1 \qquad \text{and} \qquad d_1 = P(1, 1)$$

For reference purposes we include FORTRAN mechanizations of these factorization algorithms.

Upper Triangular Cholesky Factorization

$P(N, N)$, $P = UU^{\mathrm{T}}$, $P > 0$, U Upper triangular

```
  DO 5 J = N, 2, - 1
  U(J, J) = SQRT(P(J, J))
  α = 1./U(J, J)
  DO 5 K = 1, J - 1
  U(K, J) = α * P(K, J)
  β = U(K, J)
  DO 5 I = 1, K
5 P(I, K) = P(I, K) - β * U(I, J)
  U(1, 1) = SQRT(P(1, 1))
```

Comments: The matrix U can replace P in computer storage. The lower parts of P and U are not used at all and in any case the upper part of P is destroyed by this mechanization.

Upper Triangular UDU^{T} Factorization

$P(N, N)$, $P = UDU^{\mathrm{T}}$, $P > 0$, U Unit upper triangular, D Diagonal

```
  DO 5 J = N, 2, - 1
  D(J) = P(J, J)
  α = 1./D(J)
  DO 5 K = 1, J - 1
  β = P(K, J)
  U(K, J) = α * β
  DO 5 I = 1, K
5 P(I, K) = P(I, K) - β * U(I, J)
  D(1) = P(1, 1)
```

Comments: This mechanization is such that the lower triangular portion of P is not used and U and D can share the upper triangular portion of P (the diagonal elements of U are implicitly unity). In any case the upper triangular portion of P is destroyed by this mechanization.

We include a vector stored FORTRAN mechanization of this algorithm and state it in a form with U and D replacing P. As can be seen, the code is compact and only storage for an $(N(N+1)/2)$ dimensioned vector is required. The reader interested in a vector stored version of the upper triangular square root factorization has only to exploit the commonality of the FORTRAN mechanizations of the UU^{T} and UDU^{T} algorithms.

$$P(N(N+1)/2), \qquad P = UDU^{\mathrm{T}}$$

The resulting P will have $D(J)$ stored in $P(J(J+1)/2)$ and $U(I, J)$ stored in $P(J(J-1)/2 + I)$, i.e., the D elements are on the diagonals and U is in the "upper triangular" part of P.

```
  JJ = N(N + 1)/2
  JJN = JJ
  DO 5 J = N, 2, - 1
  α = 1./P(JJ)
  KK = 0
  JJN = JJ - J                          @ JJN = next diagonal
  DO 4 K = 1, J - 1
  β = P(JJN + K)                        @ JJN + K = (K, J)
  P(JJN + K) = α * β
  DO 3 I = 1, K
3 P(KK + I) = P(KK + I) - β * P(JJN + I)
                                        @ KK + I = (I, K)
4 KK = KK + K                           @ KK = K(K - 1)/2
5 JJ = JJN                              @ JJ = J(J - 1)/2
```

Note that while this mechanization using vector storage requires essentially the same number of steps as before, the code appears somewhat more complicated. The reason for this is that matrix element indices are computed by incrementing, rather than by the more direct but time consuming double subscript–single subscript conversion formula. (Recall that the expressions to the right of the @ symbol are merely comments.)

APPENDIX III.B FORTRAN Mechanization of the Lower Triangular Cholesky Factorization

For reference purposes a lower triangular Cholesky factorization mechanization is included.

$P(N, N)$, $P = LL^{\mathrm{T}}$, $P > 0$, L Lower triangular

```
  DO 5 J = 1, N - 1
  L(J, J) = SQRT(P(J, J))
```

```
   α = 1./L(J, J)
   DO 5 K = N, J + 1, − 1
   β = L(K, J)
   L(K, J) = α * β
   DO 5 I = K, N
5  P(I, K) = P(I, K) − L(I, J) * β
   L(N, N) = SQRT(P(N, N))
```

Comments: L can share the lower part of P if desired. In any case, the lower part of P is destroyed by the algorithm.

APPENDIX III.C FORTRAN Mechanization of the UDU^T Update

The equation $\bar{U}\bar{D}\bar{U}^{\mathrm{T}} = UDU^{\mathrm{T}} + caa^{\mathrm{T}}$ is called an *update* because it can be used to include process noise into the U–D factored covariance matrix (cf. Chapter VI) or to update a U–D factored information matrix when a scalar measurement is included (cf. Agee and Turner [4]).

U, $\bar{U}$ Unit upper triangular, $U(N, N)$, $\bar{U}(N, N)$
D, $\bar{D}$ Diagonal
c is a scalar and a an N vector.

```
   DO 5 J = N, 2, − 1
   s = a(J)
   D = D(J) + c * s ** 2
   b = c/D
   β = s * b
   c = b * D(J)
   D̄(J) = D
   DO 5 I = J − 1, 1, − 1
   a(I) = a(I) − s * U(I, J)
5  Ū(I, J) = U(I, J) + β * a(I)
   D̄(1) = D(1) + c * a(1) ** 2
```

Comments: c and a are destroyed by the algorithm. $\bar{U}$ and $\bar{D}$ can replace U and D in computer storage. The diagonal elements of U are implicitly assumed to be unity and the lower triangular portion of U is implicitly assumed to be zero. $U(I, J)$ for $I \geqslant J$ are not explicitly used. The algorithm requires only trivial modification when U is given in vector form.

References

[1] Fadeev, D. K., and Fadeeva, V. N., "Computational Methods of Linear Algebra," Chapter 1. Freeman, San Francisco (1963).

Computational techniques have improved markedly since the appearance of this book. Nonetheless it is still a quite useful supplemental reference.

[2] Noble, B., "Applied Linear Algebra." Prentice-Hall, Englewood Cliffs, New Jersey (1969).

This book contains an elementary exposition of various useful notions in numerical linear algebra. It is a handy reference that supplements the linear algebra topics of this book.

[3] Wilkinson, J. H., Error analysis of direct methods of matrix inversion, *J. Assoc. Comput. Mach.* **8**, 281–330 (1961).

This early work sets the stage for much of the computer roundoff and numeric stability analyses that have appeared since.

[4] Agee, W. S., and Turner, R. H., Triangular decomposition of a positive definite matrix plus a symmetric dyad with application to Kalman filtering, White Sands Missile Range Tech. Rep. No. 38 (1972).

In this work the authors point out the value of using square root free triangularization factorizations for Kalman filtering. Their results are applicable to information matrix data processing and covariance matrix process noise inclusion. However, modification of their results (cf. Section V.3) is recommended for covariance data processing.

[5] Gill, P. E., Golub, G. H., Murray, W., and Saunders, M. A., Methods for modifying matrix factorizations, *Math. Comp.* **28**, 505–535 (1974).

Various methods for updating PD matrices are surveyed. Using orthogonal transformations these authors derive algorithms that have better numeric characteristics than the Agee–Turner factorization. The price of this increased numerical stability is an increased computational requirement.

IV

Householder Orthogonal Transformations

Square root information filtering is one of the principal topics of our estimation studies. Our SRIF mechanization depends on the Householder orthogonal transformation. In this chapter the Householder transformation will be defined and certain of its properties developed.

IV.1 Review of Orthogonal Transformations

In this section we review the concept of an orthogonal transformation and how it can be applied to the least squares problem.

Definition Matrix T, $T(m, m)$ is an *orthogonal matrix* iff $TT^{\mathrm{T}} = I$.

Properties of orthogonal matrices that are useful to know are

Property A If T_1 and T_2 are orthogonal matrices, then so is $T_1 T_2$.

Property B $T^{-1} = T^{\mathrm{T}}$ and $T^{\mathrm{T}}T = I$.

Property C For any m vector y,

$$\|Ty\| = \|y\| \tag{1.1}$$

where $\|\cdot\|$ is the Euclidean norm ($\|y\| = (y^{\mathrm{T}}y)^{1/2}$).

Property D If ν is an m vector of random variables with $E(\nu) = 0$, $E(\nu\nu^{\mathrm{T}}) = I$, then $\overline{\nu} = T\nu$ has the same properties, i.e.,

$$E(\overline{\nu}) = 0, \qquad E(\overline{\nu}\,\overline{\nu}^{\mathrm{T}}) = I \tag{1.2}$$

While this property is trivial to verify, it is a little surprising that the components of the transformed noise vector remain uncorrelated.

Properties C and D play an essential role in square root estimation.

IV.2 Application of Orthogonal Matrices to the Least Squares Problem

Recall the least squares performance functional from Chapter II:

$$J(x) = \|z - Ax\|^2, \qquad A(m, n), \qquad m \geqslant n^{\dagger}$$

and let T, $T(m, m)$ be an orthogonal matrix. Because of Property C, we can write

$$J(x) = \|T(z - Ax)\|^2 = \|(Tz) - (TA)x\|^2 \tag{2.1}$$

Now the minimum of $J(x)$, $J(x_{\mathrm{ls}})$ is independent of T and this can be exploited, i.e., T can be chosen so that (TA) has a computationally attractive form. Indeed, we shall show in Section IV.3 how T can be chosen so that

$$TA = \begin{bmatrix} R \\ 0 \end{bmatrix} \begin{matrix} \}n \\ \}m - n \end{matrix} \tag{2.2}$$

where R is an upper triangular matrix.

If Tz is partitioned,

$$Tz = \begin{bmatrix} z_1 \\ z_2 \end{bmatrix} \begin{matrix} \}n \\ \}m - n \end{matrix} \tag{2.3}$$

then $J(x)$ can be written as

$$J(x) = \|z_1 - Rx\|^2 + \|z_2\|^2 \tag{2.4‡}$$

†For example m is larger than n when the *a priori* is included as n additional equations (cf. Eq. (II.5.4)).

‡Appendix IV.B contains the back substitution method for solving a triangular system $Rx = z_1$ (cf. (2.5)) and inverting a triangular matrix.

By reducing the least squares performance functional to the form (2.4), one can see by inspection that the minimizing x_{ls} must satisfy

$$Rx = z_1 \tag{2.5}$$

and that

$$J(x_{\mathrm{ls}}) = \|z_2\|^2 \tag{2.6}$$

These results are more elegant than is the brute force construction via the normal equations. More importantly, the solution using orthogonal transformations (Eqs. (2.2), (2.3), (2.5), and (2.6)) is less susceptible to errors due to computer roundoff. Proof of this assertion is beyond the scope of our studies, and the reader is referred to Lawson and Hanson [4] and Wilkinson [5] for details. We do, however, provide in Section V.5 examples illustrating numerical failures that arise in data processing problems.

IV.3 The Householder Transformation

The Householder transformations to be discussed in this section are matrix representations corresponding to the geometric notion of a reflection. Let u be a nonzero vector and let $U_\perp$ be the plane perpendicular to u, i.e., u is the normal to the plane $U_\perp$. If y is an arbitrary vector, then referring to Fig. IV.1 we can write an analytical expression for y_{r}, the *reflection of* y in the plane $U_\perp$. Thus if

$$y = (y^{\mathrm{T}}\hat{u})\hat{u} + v \tag{3.1}$$

where $\hat{u}$ is a unit vector in the direction of u, $\hat{u} = u/\|u\|$, and v is that part of y that is orthogonal to u; then

$$y_{\mathrm{r}} = -(y^{\mathrm{T}}\hat{u})\hat{u} + v \tag{3.2}$$

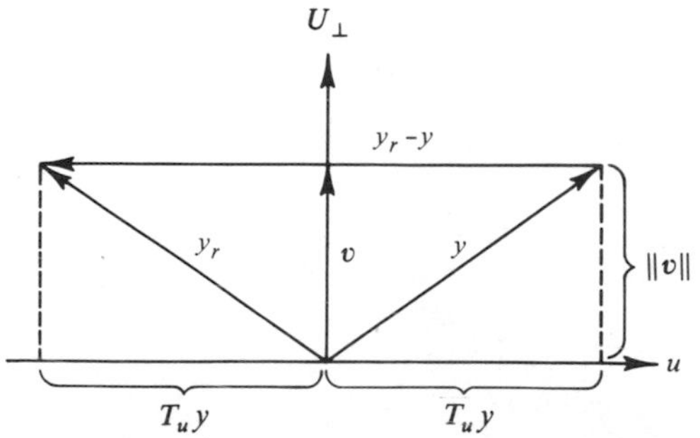

Fig. IV.1 Geometry of the Householder transformation.

Eliminating v from these two equations gives

$$y_r = y - 2(y^T\hat{u})\hat{u} = (I - \beta uu^T)y \tag{3.3}$$

where $\beta = 2/\|u\|^2 = 2/u^Tu$. The matrix $T_u = I - \beta uu^T$ is fundamental for our applications. We now enumerate properties of the *elementary Householder transformation* T_u that are useful for our estimation applications.

Property 1 $T_u = T_u^T$, T_u is *symmetric*.

Property 2 $T_u^2 = I$, T_u is *idempotent*. This property and Property 1 show that T_u is an orthogonal transformation.

Remark: One can demonstrate Property 2 algebraically by expanding T_u^2, or geometrically by referring to Fig. IV.1.

Property 3 If $u(j) = 0$, then $(T_uy)(j) = y(j)$, i.e., if the jth component of u is zero, then T_u leaves the jth component of y unchanged.

Property 4 If $u \perp y$, then $T_uy = y$.

Property 5

$$T_uy = y - \gamma u, \qquad \gamma = 2y^Tu/u^Tu \tag{3.4}$$

Property 5 is pragmatically important. Formation of T_u as a precursor to computing T_uy requires an order of magnitude more computation than does the direct evaluation of T_uy using Property 5. Further, this property shows that storage of the matrix T_u is not necessary. This point is especially important when m is large.

Triangularizing a Matrix Using Householder Transformations

We turn now to the development of a property that is basic for matrix triangularization. Given y, let us choose the direction of u so that $y_r = T_uy$ lies along e_1, i.e., $T_uy = (\sigma, 0, \ldots, 0)^T$. From Property C, Eq. (1.1), the length of y is invariant, therefore

$$\|T_uy\| = |\sigma| = (y^Ty)^{1/2}$$

The direction of u can be obtained from Property 5, Eq. (3. 4),

$$u = \text{const}(y - \sigma e_1) \tag{3.5}$$

This result leads us to:

Property 6 Let $\sigma = \mathrm{sgn}(y(1))\ (y^{\mathrm{T}}y)^{1/2}$ and define $u(1) = y(1) + \sigma$, $u(j) = y(j), j > 1$. Then $T_u y = -\sigma e_1$ and $2/u^{\mathrm{T}}u = 1/(\sigma u(1))$.

Remark: The geometric significance of (3.5) is apparent from Fig. IV.1 where it is shown that $y_r - y$ is orthogonal to $U_\perp$ and so is parallel to u.

A direct expansion of $u^{\mathrm{T}}u$ shows that $u^{\mathrm{T}}u = 2\sigma u(1)$, and the sign of σ is chosen to maximize $|u(1)|$. In the application of T_u to other vectors the term $1/(\sigma u(1))$ appears, and maximizing $|u(1)|$ avoids numerical problems. If Property 6 is applied to a matrix A (with y as the first column of A), the step that is fundamental to matrix triangularization results.

Lemma 3.1 Let $A(m, n)$ be given. Then there exists an orthogonal transformation T_u such that

$$T_u A = \begin{array}{r} 1\{ \\ m-1\left\{ \begin{array}{c} \\ \\ \end{array}\right. \end{array} \overset{\begin{array}{cc} \overbrace{1} & \overbrace{n-1} \end{array}}{\left[\begin{array}{c|c} s & \\ & \tilde{A} \\ 0 & \end{array} \right]} \tag{3.6}$$

Remark: s and $\tilde{A}$ are computed directly from A (cf. Appendix IV.A), and the matrix T_u is only implicit. Computer mechanization of the basic result (3.6) requires no additional computer storage other than that used for A, because A and $\tilde{A}$ share common storage and T_u is not explicitly computed.

Repeated application of Property 3 and the lemma gives the matrix triangularization algorithm:

Theorem 3.1 (Matrix Triangularization) Let $A(m, n) = A_1$ and for each j choose an elementary Householder matrix T_j so that

$$T_j A_j = \overset{\begin{array}{cc} \overbrace{1} & \overbrace{n-j} \end{array}}{\begin{bmatrix} s_j & a_j^{\mathrm{T}} \\ 0 & A_{j+1} \end{bmatrix}} \begin{array}{l} \}\ 1 \\ \}\ m-1 \end{array} \qquad j = 1, \ldots, k \quad (k \leqslant n) \tag{3.7}$$

Then

$$T = \begin{bmatrix} I_{k-1} & 0 \\ 0 & T_k \end{bmatrix} \cdots \begin{bmatrix} I_1 & 0 \\ 0 & T_2 \end{bmatrix} T_1 \tag{3.8}$$

is an orthogonal transformation and

$$TA = \begin{bmatrix} s_1 & & & a_1^{\mathrm{T}} \\ & s_2 & & a_2^{\mathrm{T}} \\ & & \ddots & \vdots \\ 0 & & s_k & a_k^{\mathrm{T}} \\ & & & A_{k+1} \end{bmatrix} \tag{3.9}$$

It is important to note that the triangularization algorithm (3.7) does not require computation or storage of the orthogonal matrix T. None of our intended estimation applications requires T and for this reason computer implementation involves only (3.7). This process is developed in greater detail in Appendix IV.A.

This theorem is the cornerstone of the square root information algorithms that are developed in this book. Observe how the A_j are implicitly defined by recursion (3.7). Note also that the jth step of this recursion also defines the scalar s_j and the $n - j$ component row vector a_j^{T}. In Appendix IV.A the implicit relations for s_j, a_j^{T}, and A_{j+1} are made explicit and a FORTRAN mechanization is given in Appendix VII.B.

Turning our attention to Eq. (3.9), we observe that the jth transformation zeros the jth column of A, below row j; determines the elements of the jth row that are to the right of the jth column, with s_j as the jth diagonal element; and does not alter, in any way, the rows above the jth. Thus the algorithm of the theorem can be used to orthogonally transform a matrix to a partially triangular form.

In our investigations we will have occasion to use this theorem (cf. Fig. IV.2) with

(a) $k = m < n$; in which case the lower triangular portion of the transformed matrix will contain zeros.

(b) $m = n = k$; which corresponds to the transformation of A to an upper triangular form.

(c) $m > n = k$; in this case A is transformed to an upper triangular form, but the block of zeros below the diagonal is not a triangular array.

(d) $m > n > k$; corresponds to the partial triangularization depicted in (3.9).

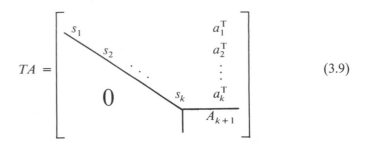

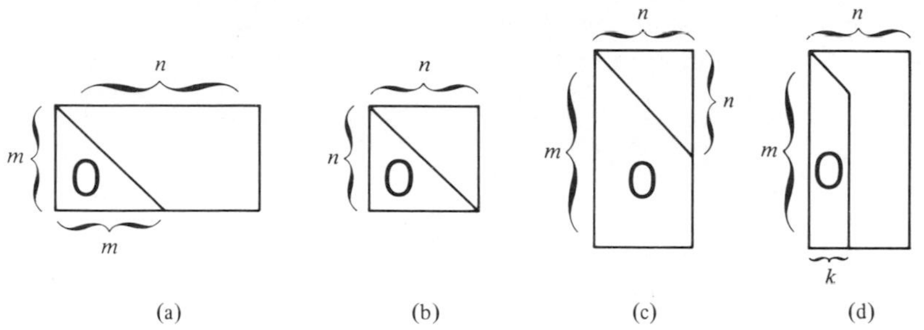

Fig. IV.2 Schematic visualization of the various matrix triangularization applications.

APPENDIX IV.A Annihilating the First Column of a Matrix Using the Householder Transformation

Let $A(m, n)$ be given. Then the basic operation of the triangularization process is the construction of the scalar s and $\tilde{A}(m, n-1)$ so that

$$T_u A = \begin{bmatrix} s & \\ & \tilde{A} \\ 0 & \end{bmatrix} \tag{A.1}$$

The algorithms for computing s and $\tilde{A}$ are direct applications of Property 6 and Lemma 3.1:

$$s = -\operatorname{sgn}(A(1, 1))\left(\sum_{i=1}^{m} [A(i, 1)]^2\right)^{1/2} \tag{A.2}$$

$$u(1) = A(1, 1) - s \tag{A.3}$$

$$u(i) = A(i, 1), \qquad i = 2, \ldots, m \tag{A.4}$$

$$\beta = 1/(su(1)) \tag{A.5}$$

For $j = 2, \ldots, n$, evaluate Eqs. (A.6)–(A.7) (apply T to the successive columns of A):

$$\gamma := \beta \cdot \sum_{i=1}^{m} u(i)A(i, j) \tag{A.6}$$

$$\tilde{A}(i, j-1) = A(i, j) + \gamma u(i), \qquad i = 1, \ldots, m \tag{A.7}$$

Appendix VII.B contains a FORTRAN mechanization of the orthogonal reduction of a rectangular matrix to a triangular form; there we omit $\tilde{A}$ and replace (A.7) by

$$A(i,j) := A(i,j) + \gamma\, u(i)$$

APPENDIX IV.B Solution of the Triangular System Rx = y and Inversion of a Triangular Matrix

In our applications of square root estimation we shall frequently be interested in solving

$$Rx = z \tag{B.1}$$

where $R(n, n)$ is upper triangular (and assumed nonsingular). If only the solution x is needed, one should not circuitously compute R^{-1}, and then $R^{-1}z$. The following back substitution algorithm computes x directly.

For $j = n, n-1, \ldots, 1$, compute

$$x(j) = \left(z(j) - \sum_{k=j+1}^{n} R(j,k)x(k)\right) \Big/ R(j,j) \tag{B.2}$$

Note that the algorithm is about as complicated as a matrix multiplication. One can, if desired, write the $x(j)$ over the $z(j)$. This is convenient for some applications.

If one has specific need for the inverse $R^{-1} = U$, it can be computed using a bordering algorithm that is in keeping with the spirit of recursive computation, which is the theme of this text. The following identity for triangular matrices is easily verified:

$$\begin{bmatrix} R_j & y \\ 0 & \sigma_{j+1} \end{bmatrix}^{-1} = \begin{bmatrix} R_j^{-1} & -R_j^{-1} y \sigma_{j+1}^{-1} \\ 0 & \sigma_{j+1}^{-1} \end{bmatrix} = R_{j+1}^{-1} \tag{B.3}$$

This relation (cf. Householder (1964), pp. 125–126) enables one to compute recursively the inverse of an ever larger matrix, i.e., if $R_j^{-1} = U_j$, where R_j is the upper left $j \times j$ partition of R, then

$$U_{j+1} = \begin{bmatrix} U_j & -U_j(R(1,j+1), \ldots, R(j,j+1))^{\mathrm{T}} \sigma_{j+1} \\ 0 & \sigma_{j+1} \end{bmatrix} \tag{B.4}$$

where $\sigma_{j+1} = 1/R(j+1, j+1)$.

This result is readily transformed into algorithmic form by setting $U = R^{-1}$,

$$U(1, 1) = 1/R(1, 1) \tag{B.5}$$

For $j = 2, \ldots, n$ evaluate (B.6) and (B.7),

$$U(j, j) = 1/R(j, j) \tag{B.6}$$

$$U(k, j) = -\left(\sum_{l=k}^{j-1} U(k, l)R(l, j)\right)U(j, j), \qquad k = 1, \ldots, j-1 \tag{B.7}$$

Remarks: (1) R^{-1} is computed a column at a time so that it can be expanded if one wants to consider augmenting the list of parameters.

(2) The inversion is such that if desired one can store the inverse in the locations previously assigned to R.

A FORTRAN mechanization of Eqs. (B.5)–(B.7) follows.

Comment: $U = R^{-1}$, where R is upper triangular, and U can replace R if desired.

```
    U(1, 1) = 1./R(1, 1)
    DO 20 J = 2, N
    U(J, J) = 1./R(J, J)
    JM1 = J - 1
    DO 20 K = 1, JM1
    SUM = 0.
    DO 10 I = K, JM1
10  SUM = SUM - U(K, I) * R(I, J)
20  U(K, J) = SUM * U(J, J)
```

In applications where computer storage is at a premium and/or where storage of the triangular matrix as a doubly dimensioned array is not feasible, one may store the upper triangular portions of R and U as vectors (cf. Appendix VII.B and Section III.2). It is still possible to compute R^{-1} based on Eqs. (B.5)–(B.7), but now one also calculates a variety of indices (matrix subscripts). Since computers generally add and subtract more rapidly than they multiply and divide there is a strong motivation to arrange the index computations in a recursive incremental form. To illustrate our point, and because matrix inversion of triangular, vector-stored matrices is not uncommon in SRIF applications (cf. Chapter VII) we include a FORTRAN mechanization of this algorithm.

Comments: $U = R^{-1}$, where R is an upper triangular matrix; U and R are stored as vectors of dimension $N(N + 1)/2$; and U can replace R if desired.

```
   U(1) = 1./R(1)              @ R(1) = R(1, 1)
   JJ = 1
   DO 20 J = 2, N
   JJ‾ = JJ                    @ JJ‾ = J(J − 1)/2 = (J − 1, J − 1)
   JJ = JJ + J                 @ JJ = J(J + 1)/2 = (J, J)
   U(JJ) = 1./R(JJ)
   JM1 = J − 1
   KK = 0
   DO 20 K = 1, JM1
   KK = KK + K                 @ KK = K(K + 1)/2
   KI = KK
   SUM = 0.
   DO 10 I = K, JM1
   SUM = SUM − U(KI) * R(JJ‾ + I)
                               @ KI = (K, I), JJ‾ + I = (I, J)
10 KI = KI + I                 @ KI = (K, I + 1)
20 U(JJ‾ + K) = SUM * U(JJ)    @ JJ‾ + K = (K, J)
```

This mechanization is a direct consequence of the previous one. The conversion to triangular form is accomplished by identifying $J(J - 1)/2 + I$ with (I, J) (cf. Section III.3). Note that recursively incrementing KK, JJ, and KI saves computation.

References

[1] Golub, G. H., Numerical methods for solving linear least squares problems, *Num. Math.* **7**, 206–216 (1965).

Tutorial, well written, and full of useful information.

[2] Businger, P., and Golub, G. H., Linear least squares solutions by Householder transformations, *Num. Math.* **7**, 269–276 (1965).

This reference is a computer program.

[3] Lawson, C. L., and Hanson, R. J., Extensions and applications of the Householder algorithm for solving linear least squares problems, *Math. Comp.* **23**, No. 108, 787–812 (1969).

Well written and full of useful information.

[4] Lawson, C. L., and Hanson, R. J., "Solving Least Squares Problems." Prentice-Hall, Englewood Cliffs, New Jersey (1974).

This book gives an in-depth, comprehensive treatment of the numerical techniques that apply to ill-conditioned least squares problems.

[5] Wilkinson, J., Error analysis of transformations based on the use of matrices of the form I-$2ww^T$, *in* "Errors in Digital Computation" (L. B. Rall, ed.), Volume II, pp. 77–101. Wiley, New York (1965).

V

Sequential Square Root Data Processing

V.1 Introduction

The mathematical machinery developed in the last chapter on Householder orthogonal transformations makes it a simple task to develop the *SRIF* (square root information filter) data processing algorithm. In Chapter II we derived the classical information filter algorithm (sequential processing of the normal equations), the Kalman filter data processing algorithm, and the square root mechanization introduced by Potter. The SRIF algorithm, to be developed in Section V.2, is an alternative to these other algorithms, and like the Potter covariance square root it is reputed to be more accurate[†] and stable[‡] than the conventional non-square root algorithms. It is of interest to note that although the Potter and the SRIF square root methods have been observed to give comparable accuracy, only the SRIF, mechanized with Householder orthogonal transformations, has been rigorously analyzed (cf. Lawson and Hanson (1974)). We note, however, that Bierman's work [4] shows that Potter's algorithm is in fact numerically sound. There it is shown that the Potter result can be derived by applying an elementary Householder orthogonal transformation to an augmented matrix.

To round out our study of data processing algorithms we will, in Section V.3, develop our U–D factorization of the Kalman update algorithm ($P = UDU^{\mathrm{T}}$, with U unit upper triangular and D diagonal). This algorithm is considered to be of square root type since $UD^{1/2}$ is a covariance square root that has been studied by Carlson [1]. The U–D updating algorithm has much to recommend it, viz., it appears to have the accuracy characteristic of square root algorithms (cf. Section V.5), it does

[†]Theoretically, all these algorithms give identical results when "perfect" arithmetic is performed. More *accurate*, here, means less susceptible to the effects of computer roundoff errors.

[‡]By *stable* we mean that accumulated roundoff errors will not cause the algorithm to diverge.

not involve scalar square roots and thus qualifies for use in real-time applications that involve minicomputers with limited capability, it is considerably more efficient than are the Potter and Carlson square root covariance algorithms and is in fact almost as efficient as the (unstable) Kalman algorithm (cf. Section V.4), and lastly the triangular structure of the algorithm enables one to consider simultaneously parameter estimation models of various dimensions.

The topics to be discussed in this chapter are as follows. In Section V.2. Golub's SRIF data processing algorithm† is derived and some of its properties are highlighted. Section V.3 is devoted to the development and discussion of our U–D algorithm. Detailed arithmetic operation counts of the factorization algorithms are tabulated, in Section V.4, along with those of the conventional covariance and information algorithms. Extensive comparisons are made, and it is shown that there is no algorithm that is best under all circumstances. The effects of computer roundoff are analyzed, by example, in Section V.5. Prototype examples are considered that demonstrate how the conventional algorithms can fail and how the factorization algorithms overcome these problems.

V.2 The SRIF Data Processing Algorithm

The SRIF data processing algorithm to be developed in this section appears to be complicated and cumbersome to implement because it uses orthogonal transformations that are only implicitly defined. The complexity is only apparent and is due, in the main, to the representation of the algorithm using matrix operations. FORTRAN realization of $[R^{\mathrm{T}}\ 0]^{\mathrm{T}} = \pi_i(I - \beta_i u_i u_i^{\mathrm{T}})A$ (cf. Chapter IV) requires little more storage and code (cf. Appendix VII.A) than do the Kalman, Potter, and U–D algorithms.

To begin our development suppose that we have an *a priori information array* $[\tilde{R}\ \tilde{z}]$, which corresponds to the *data equation* $\tilde{z} = \tilde{R}x + \tilde{\nu}$, where $\tilde{\nu}$ is assumed to have zero mean and unity covariance, and $\tilde{R}$ is square. In the case where $\tilde{R}$ is nonsingular the *a priori* covariance and estimate $[\tilde{P}\ \tilde{x}]$ are related to $\tilde{R}$ and $\tilde{z}$ by

$$\tilde{x} = \tilde{R}^{-1}\tilde{z}, \qquad \tilde{P} = \tilde{R}^{-1}\tilde{R}^{-\mathrm{T}} \tag{2.1}$$

Equation (2.1) is a direct consequence of the data equation. This follows from the least squares estimate and covariance results of Chapter II, i.e.,

$$\tilde{x} = (\tilde{R}^{\mathrm{T}}\tilde{R})^{-1}\tilde{R}^{\mathrm{T}}\tilde{z} = \tilde{R}^{-1}\tilde{z} \qquad \text{(cf. Eq. (II. 2.4))}$$

$$\tilde{P} = (\tilde{R}^{\mathrm{T}}\tilde{R})^{-1} = \tilde{R}^{-1}\tilde{R}^{-\mathrm{T}} \qquad \text{(cf. Eq. (II. 3.4))}$$

†The sequential SRIF formulation which is crucial for our estimation applications is, however, due to Hanson and Lawson (1969).

Conversely, as was pointed out in Chapter II, one can construct a data equation from the covariance–estimate pair $[\tilde{P}\ \tilde{x}]$, i.e.,

$$\tilde{P}^{-1/2}\tilde{x} = \tilde{P}^{-1/2}x + \tilde{P}^{-1/2}(\tilde{x} - x)$$

where $\tilde{P}^{-1/2}$ is a square root of $\tilde{P}^{-1}$.

Remarks: (1) The name "square root information filter" (SRIF) is due to the observation (cf. (II.3.4)) that the information matrix P^{-1} equals $R^{\mathrm{T}}R$ and thus R^{T} is a square root of P^{-1}.

(2) The data equation corresponding to the covariance–estimate pair is not unique, because if $Rx = z - \nu$ with ν having unity covariance and T is an orthogonal matrix, then $(TR)x = Tz - T\nu$, and $T\nu$ still has unity covariance (cf. Chapter IV). Thus the information array $[R\ z]$ is equivalent to the information array $[TR\ Tz]$. We shall make use of this property in Chapter VI by choosing the transformation so that TR is partially triangular.

We are interested here in constructing the least squares solution to the *a priori data equation* $\tilde{z} = \tilde{R}x + \tilde{\nu}$, and the new measurements $z = Ax + \nu$. In Chapter II it was shown that the least squares solution could be obtained by forming the normal equations; here we construct the least squares solution by applying the method of orthogonal transformations described in Section IV.2.

Finding the least squares solutions to

$$\begin{bmatrix} \tilde{R} \\ A \end{bmatrix} x = \begin{bmatrix} \tilde{z} \\ z \end{bmatrix} - \begin{bmatrix} \tilde{\nu} \\ \nu \end{bmatrix} \tag{2.2}$$

was seen to be equivalent to finding the least squares solution to

$$T\begin{bmatrix} \tilde{R} \\ A \end{bmatrix} x = T\begin{bmatrix} \tilde{z} \\ z \end{bmatrix} - T\begin{bmatrix} \tilde{\nu} \\ \nu \end{bmatrix} \tag{2.3}$$

where T is an orthogonal transformation. In Chapter IV it was shown that T can be chosen so that

$$T\begin{bmatrix} \tilde{R} \\ A \end{bmatrix} = \begin{bmatrix} \hat{R} \\ 0 \end{bmatrix} \tag{2.4}$$

where $\hat{R}$ is an upper triangular matrix. If we set

$$T\begin{bmatrix} \tilde{z} \\ z \end{bmatrix} = \begin{bmatrix} \hat{z} \\ e \end{bmatrix}; \qquad T\begin{bmatrix} \tilde{\nu} \\ \nu \end{bmatrix} = \begin{bmatrix} \hat{\nu} \\ \nu_{\mathrm{e}} \end{bmatrix}$$

then (2.3) reduces to

$$\hat{R}x = \hat{z} - \hat{\nu} \tag{2.5}$$

$$0 = e - \nu_e \tag{2.6}$$

Equation (2.5) is in the form of a data equation. This is important because this equation can now act as the *a priori* data equation that is to be combined with the next set of observations and the procedure repeated. In this way we have developed a recursive algorithm to process observations sequentially.

The e in Eq. (2.6) has an interpretation. It is the error in the least squares fit. To see this recall that

$$J(x) = \left\| \begin{bmatrix} \tilde{R} \\ A \end{bmatrix} x - \begin{bmatrix} \tilde{z} \\ z \end{bmatrix} \right\|^2$$

$$= \|\hat{R}x - \hat{z}\|^2 + \|e\|^2 \qquad \text{(cf. Eq. (IV.2.4))} \tag{2.7}$$

Thus $\|e\|^2$ is the SOS (*sum of squares*) residual error corresponding to the least squares solution of (II.2.1).

To arrange results (2.2)–(2.6) in the form of a recursive algorithm we note that the key operation explicitly involves only the augmented information array

$$\begin{bmatrix} \tilde{R} & \tilde{z} \\ A & z \end{bmatrix}$$

and the resultant information array $[\hat{R}\ \hat{z}]$. Thus the algorithm can be succinctly expressed as the

SRIF Data Processing Algorithm

Let the *a priori* information array be $[\tilde{R}_0\ \tilde{z}_0]$, and let the measurements be $\{z_i = A_i x + \nu_i\}_{i=1}^m$, where each z_i is a vector. Then the least squares estimate can be generated recursively via

$$T_j \begin{bmatrix} \hat{R}_{j-1} & \hat{z}_{j-1} \\ A_j & z_j \end{bmatrix} = \begin{bmatrix} \hat{R}_j & \hat{z}_j \\ 0 & e_j \end{bmatrix}, \qquad j = 1, \ldots, m \tag{2.8}$$

where $\left[\hat{R}_0\ \hat{z}_0\right] = \left[\tilde{R}_0\ \tilde{z}_0\right]$ and T_j, orthogonal, is chosen to triangularize the augmented matrix

$$\begin{bmatrix} \hat{R}_{j-1} \\ A_j \end{bmatrix}$$

Note that $\hat{x}_j = \hat{R}_j^{-1} z_j$ is the least squares estimate corresponding to the *a priori* information and the measurement set $\{z_i\}_{i=1}^{j}$, and that $\hat{P}_j = \hat{R}_j^{-1}\hat{R}_j^{-\mathrm{T}}$ is the covariance of this estimate (cf. Eq. (2.1)). Because $\hat{R}_j$ is triangular, the calculation of $\hat{x}_j$ and/or $\hat{P}_j$ is easily accomplished using back substitution methods (cf. Chapter IV). The vector $[^{\hat{z}_j}_{e_j}]$ is the result of applying T_j to the vector $[^{\hat{z}_{j-1}}_{z_j}]$; e_j is an error term whose significance will soon be discussed (cf. Eq. (2.14)).

One might ask whether a different answer might result if, instead of (2.8), one considered the entire system

$$\begin{bmatrix} \hat{R}_0 & \hat{z}_0 \\ A_1 & z_1 \\ \vdots & \vdots \\ A_m & z_m \end{bmatrix} \tag{2.9}$$

The answer is no; that is, the same answer would result. To see this, we show that a sequence of orthogonal transformations and equation permutations can be chosen, which when applied to the total system (2.9), gives the result (2.8). The sequence of transformations and permutations starts with

$$\left[\begin{array}{c|c} T_1 & 0 \\ \hline 0 & I \end{array}\right] \left[\begin{array}{cc} \hat{R}_0 & \hat{z}_0 \\ A_1 & z_1 \\ \hline A_2 & z_2 \\ \vdots & \vdots \\ A_m & z_m \end{array}\right] = \left[\begin{array}{cc} \hat{R}_1 & \hat{z} \\ 0 & e_1 \\ \hline A_2 & z_2 \\ \vdots & \vdots \\ A_m & z_m \end{array}\right] \tag{2.10}$$

Notice that the elements with subscripts greater than or equal to two remain unchanged. The matrix row $[0\ e_1]$ can be moved to the bottom because this corresponds to permuting the associated equations. Next, apply an orthogonal transformation to the permuted right side of (2.10):

$$\begin{bmatrix} T_2 & 0 \\ 0 & I \end{bmatrix} \begin{bmatrix} \hat{R}_1 & \hat{z}_1 \\ A_2 & z_2 \\ \hline A_3 & z_3 \\ \vdots & \vdots \\ A_m & z_m \\ 0 & e_1 \end{bmatrix} = \begin{bmatrix} \hat{R}_2 & \hat{z}_2 \\ 0 & e_2 \\ \hline A_3 & z_3 \\ \vdots & \vdots \\ A_m & z_m \\ 0 & e_1 \end{bmatrix} \tag{2.11}$$

The row matrix $[0 \;\; e_2]$ is moved to the bottom and the procedure repeated, a total of m times, the final result being

$$\begin{bmatrix} T_m & 0 \\ 0 & I \end{bmatrix} \begin{bmatrix} \hat{R}_{m-1} & \hat{z}_{m-1} \\ A_m & z_m \\ \hline 0 & e_1 \\ \vdots & \vdots \\ 0 & e_{m-1} \end{bmatrix} = \begin{bmatrix} \hat{R}_m & \hat{z}_m \\ 0 & e_m \\ \hline 0 & e_1 \\ \vdots & \vdots \\ 0 & e_{m-1} \end{bmatrix} \tag{2.12}$$

Equations (2.10)–(2.12) prove that we obtain the same result whether or not the observations are segmented into batches.

A by-product of the orthogonal matrix method depicted by (2.9)–(2.12) is the observation that one can compute the accumulated residual SOS without explicitly computing the estimate. This is useful because when new data $z_{j+1} = A_{j+1}x + v_{j+1}$ are incorporated the estimate changes, and one would otherwise have to recalculate the past residuals, i.e.,

$$\sum_{i=1}^{j} \|z_i - A_i\hat{x}_j\|^2$$

would have to be replaced by

$$\sum_{i=1}^{j} \|z_i - A_i\hat{x}_{j+1}\|^2$$

One can obtain this conclusion, as well as recursion (2.8), from the following self-explanatory string of representations for the least squares functional:

$$
\begin{aligned}
J_1(x) &\triangleq \|\hat{R}_0 x - \hat{z}_0\|^2 + \sum_{j=1}^{m} \|A_j x - z_j\|^2 \\
&= \left\| \begin{bmatrix} \hat{R}_0 \\ A_1 \end{bmatrix} x - \begin{bmatrix} \hat{z}_0 \\ z_1 \end{bmatrix} \right\|^2 + \sum_{j=2}^{m} \|A_j x - z_j\|^2 \\
&= \|e_0\|^2 + \|\hat{R}_1 x - \hat{z}_1\|^2 + \sum_{j=2}^{m} \|A_j x - z_j\|^2 \qquad \text{(cf. Eq. (2.8))} \\
&= \|e_0\|^2 + \left\| \begin{bmatrix} \hat{R}_1 \\ A_2 \end{bmatrix} x - \begin{bmatrix} \hat{z}_1 \\ z_2 \end{bmatrix} \right\|^2 + \sum_{j=3}^{m} \|A_j x - z_j\|^2 \\
&= \|e_0\|^2 + \|e_1\|^2 + \|\hat{R}_2 x - \hat{z}_2\|^2 + \sum_{j=3}^{m} \|A_j x - z_j\|^2 \\
&\ \ \vdots \\
&= \sum_{i=0}^{k} \|e_i\|^2 + \|\hat{R}_{k+1} x - \hat{z}_{k+1}\|^2 + \sum_{j=k+2}^{m} \|A_j x - z_j\|^2 \\
&= \sum_{i=0}^{k} \|e_i\|^2 + \left\| \begin{bmatrix} \hat{R}_{k+1} \\ A_{k+2} \end{bmatrix} x - \begin{bmatrix} \hat{z}_{k+1} \\ z_{k+2} \end{bmatrix} \right\|^2 + \sum_{j=k+3}^{m} \|A_j x - z_j\|^2 \\
&\ \ \vdots \\
&= \sum_{i=0}^{m-1} \|e_i\|^2 + \|\hat{R}_m x - \hat{z}_m\|^2
\end{aligned}
\tag{2.13}
$$

The reader would be well advised to study this string of equations; they are the essence of the SRIF. From these fundamental relations we obtain the useful result

$$
\sum_{j=0}^{k-1} \|e_j\|^2 = \|\hat{R}_0 \hat{x}_k - \hat{z}_0\|^2 + \sum_{j=1}^{k} \|A_j \hat{x}_k - z_j\|^2 \tag{2.14}
$$

Note specifically that none of the e_j on the left of (2.14) are changed when data at time $k + 1$ are added. The significance of this point is that one does not need to use the past data (z's and A's) to compute the residual SOS corresponding to the most recent estimate.

The Kalman data processing algorithms and the normal equation method, which were discussed in Chapter II, do not share this property. In problems involving data fitting it is therefore often necessary to recompute all the residuals each time a new batch of observations is included. Such residual recomputation may involve reading stored observation matrices, and in the case of the normal equation processor it requires generation of solutions. Our discussion indicates that engineers and scientists involved in residual analysis of large sets of data can exploit this aspect of the SRIF algorithm.†

The derivation leading to (2.13) is very attractive because it makes clear why we deal with information arrays of the form $[R \ \ z]$, how recursion (2.8) comes about; and why $\Sigma\|e_j\|^2$ represents the sum of squares of the residuals.

Remark: As mentioned in the introduction, the recursive algorithm (2.8) is numerically stable, and is more accurate than the method of normal equations described in Chapter II (Appendix II.C).

One should not infer from (2.10)–(2.12) that processing all the observations simultaneously involves the same amount of computation as processing the observations sequentially. Indeed, the contrary is true; triangularizing the total system (2.9) requires less computation than sequentially triangularizing the m smaller matrices. For example, one needs N_x scalar square roots and N_x divisions for *each* batch, N_x being the dimension of x. It is interesting to observe that there is no penalty for subdividing the measurement vector when the method of normal equations is used.‡ Indeed, it should be noted that of all the data processing algorithms that are considered, only the SRIF is penalized (computationally) for subdividing the measurement vector. There is thus a tendency not to partition the measurement vector and to set aside a block of computer storage, of maximum dimension, to accommodate incoming measurements of the largest possible size.

An Aside: A word of caution regarding the choice of batch size is in order. The penalty for using smaller batches should be balanced against the storage required to handle large-dimensioned observation batches. For example, in orbit determination problems, (involving from 6 to 100 parameters) there is rarely a significant computational savings achieved by using batch sizes greater than 20; on the other hand, dimensioning software programs to accommodate very large observation batches sometimes leads to severe "core time product" penalties. When N_x is small, say $N_x \leqslant 10$, computing costs are generally insignificant. In such cases one

†For example, Jet Propulsion Laboratory personnel have included this capability into the JPL Orbit Determination Program.

‡Note that both the SRIF and information filter only implicitly produce estimates, and we are not at this moment discussing the computation of estimates and/or covariances at the subdivision points.

might even include vector data one component at a time with a negligible cost penalty. For example, scalar processing 1000 measurements of a six-parameter system, using double precision arithmetic, requires less than five seconds of CPU time on JPL's UNIVAC 1108 computer. ■

As mentioned earlier the orthogonal transformation T_j in (2.8) is not explictly needed; only the result of applying the transformation is needed. This point is exploited in the construction of the triangularized matrices (cf. Chapter IV and Appendix VII.A). Dr. H. Lass, of the Jet Propulsion Laboratory, carried this idea a step further and proposed† calculating the triangular matrix directly (a row at a time) without explictly involving any orthogonal transformations. His idea corresponds to computing the rows of the upper triangular R matrix such that $A^{\mathrm{T}}A = R^{\mathrm{T}}R$, i.e., R is a Cholesky factor of $A^{\mathrm{T}}A$ (cf. Chapter III). Although the method is more efficient and amenable to hand computation than the method using Householder transformations, it is less accurate.‡ The method of direct triangularization was not included in the comparison of algorithms to be discussed in Section V.4 even though it may be more accurate and numerically stable than are the Kalman filter algorithms.

The SRIF algorithm derived in this section plays a central role in our treatment of filtering and smoothing in the presence of process noise, and in the error analysis of the effects of various types of mismodeling.

V.3 Data Processing Using the U–D Covariance Factorization

As mentioned earlier, problems that involve frequent estimates are more efficiently handled using Kalman's sequential covariance estimation formulation (cf. Chapter II). In that chapter we described Potter's square root formulation of the Kalman algorithm, and here we discuss an alternative, triangular factorization (cf. Bierman (1975)). The formulation we have in mind is of interest because, compared with Potter's algorithm, it offers both computational savings and reductions in computer storage requirements. Specifically, we develop recursions for U and D, where $P = UDU^{\mathrm{T}}$ with U unit upper triangular and D diagonal. Computing with triangular matrices involves fewer arithmetic operations, and this factorization avoids time-consuming scalar square roots. The algorithm to be presented was motivated by the work of Agee and Turner (1972) but has been modified, using an observation made by Carlson [1], that avoids potentially unstable numerical differencing.

†Private communication.

‡This method, and its numerical pitfalls, are discussed in Section V.5.

Carlson's work is based on a triangular square root of P, and is related in an obvious way to the U–D decomposition which is discussed here. The recursions to be derived for U and D are equivalent to Carlson's result for the matrix square root $P^{1/2} = UD^{1/2} = U\,\mathrm{Diag}(\sqrt{d_1}, \ldots, \sqrt{d_n}),\ n = N_x$. Indeed, perusal of the computer implementations of the U–D and $P^{1/2}$ algorithms[†] (cf. Appendix V.A) shows that they are almost identical, except that the latter algorithm requires n scalar square roots and n extra divisions for each scalar observation that is to be processed. These differences are minor unless (a) n is very large, (b) there are many measurements, or (c) the algorithm is being applied in real time and the computer used is primitive, having slow divide and square root operations. In such cases the U–D result is certainly to be preferred.

It turns out that the measurement-updating algorithm for the U–D factors, to be given in Theorem V.3.1, is related to the Givens orthogonal transformation, which is famous for its numerical stability (cf. Gentlemen (1973)). Gentlemen's analyses show that the U–D measurement-updating process has numeric accuracy and stability that compares with that obtained using Householder orthogonal transformations. Extensive computer simulations, documented in the Ph.D. thesis of C. Thornton (1976), corroborate the outstanding numeric characteristics of the algorithm.

The results of Chapter II indicate how one can combine an *a priori* estimate and error covariance $\tilde{x}$ and $\tilde{P}$, respectively, with a scalar observation $z = a^{\mathrm{T}}x + \nu$ to construct an updated estimate and error covariance $\hat{x}$ and $\hat{P}$. The relevant equations are

$$K = \tilde{P}a/\alpha, \qquad \alpha = a^{\mathrm{T}}\tilde{P}a + r \qquad \text{(cf. Eq. (II.6.12))}^{\ddagger}$$

$$\hat{x} = \tilde{x} + K(z - a^{\mathrm{T}}\tilde{x}) \qquad \text{(cf. Eq. (II.6.4))}$$

$$\hat{P} = \tilde{P} - Ka\tilde{P} \qquad \text{(cf. Eq. (II.6.13))}$$

The main result of this section is the following alternative representation of this result.

Theorem V.3.1 (UDU^{T} Estimate–Covariance Updating Algorithm) Suppose that an *a priori* estimate and covariance given by $\tilde{x}$ and $\tilde{P} = \tilde{U}\tilde{D}\tilde{U}^{\mathrm{T}}$, with $\tilde{U}$ unit upper triangular and $\tilde{D}$ diagonal, is to be combined

[†]Our FORTRAN mechanization is slightly more efficient than that proposed by Carlson.

[‡]In Chapter II we used D instead of α; in this section, however, D has a different meaning. Note also that in this section we violate our convention of normalizing the measurement noise and denote it by r $(= E(\nu^2))$. The reason for this is that covariance algorithms are most often formulated in an unnormalized form.

with the scalar observation $z = a^{\mathrm{T}}x + \nu$, $E(\nu) = 0$, $E(\nu^2) = r$. The Kalman gain K and the updated covariance factors U and D can be obtained from the following algorithm:

$$\left.\begin{aligned} f &= \tilde{U}^{\mathrm{T}}a, \qquad f^{\mathrm{T}} = (f_1 \cdots f_n) \\ v &= \tilde{D}f, \qquad v_i = \tilde{d}_i f_i \qquad i = 1 \cdots n \end{aligned}\right\} \tag{3.1}$$

$$\hat{d}_1 = \tilde{d}_1 r/\alpha_1, \qquad \alpha_1 = r + v_1 f_1; \qquad K_2^{\mathrm{T}} = \Big(v_1 \; \overbrace{0 \cdots 0}^{n-1}\Big) \tag{3.2}$$

For $j = 2, \ldots, n$ recursively cycle through Eqs. (3.3)–(3.6):

$$\alpha_j = \alpha_{j-1} + v_j f_j \tag{3.3}$$

$$\hat{d}_j = \tilde{d}_j \alpha_{j-1}/\alpha_j \tag{3.4}$$

$$\hat{u}_j = \tilde{u}_j + \lambda_j K_j, \qquad \lambda_j = -f_j/\alpha_{j-1} \tag{3.5}$$

$$K_{j+1} = K_j + v_j \tilde{u}_j \tag{3.6}$$

where

$$\tilde{U} = [\tilde{u}_1 \cdots \tilde{u}_n], \qquad \hat{U} = [\hat{u}_1 \cdots \hat{u}_n]$$

and the Kalman gain is given by

$$K = K_{n+1}/\alpha_n \tag{3.7}$$

Notation: The components of subscripted vectors are given as parenthetical arguments, viz.,

$$K_j^{\mathrm{T}} = (K_j(1) \cdots K_j(n)), \qquad u_j^{\mathrm{T}} = (u_j(1) \cdots u_j(n))$$

To be consistent with the notation used elsewhere in this chapter we subscript the components of those vectors that do not themselves have subscripts, viz., $f^{\mathrm{T}} = (f_1 \cdots f_n)$ and $v^{\mathrm{T}} = (v_1 \cdots v_n)$. $\tilde{u}_j$ and $\hat{u}_j$ are the jth columns of matrices $\tilde{U}$ and $\hat{U}$, respectively.

Remarks: Before going on to the proof of this result we point out that despite its somewhat awkward appearance, when compared with, say, the simpler representations (II.6.12) and (II.6.13), our U–D algorithm is ideally suited to computer implementation (cf. Appendix V.A). Furthermore, the algorithm involves very nearly the same number of operations as the conventional covariance update (II.6.13) (cf. Section V.4) and is numerically stable (cf. Section V.5). The stability is due to the avoidance of numerical differencing in the computation of D (compare Eqs. (3.4) and the Agee–Turner result (III.4.2)). This latter equation has numerical stability properties that are similar to those of the conventional covariance update (II.6.13).

Proof: We begin by expressing the Kalman update equation (II.6.13) in terms of the vectors f and v:

$$\hat{U}\hat{D}\hat{U}^{\mathrm{T}} = \tilde{U}\Big[\tilde{D} - \alpha^{-1}(\tilde{D}\tilde{U}^{\mathrm{T}}a)(\tilde{D}\tilde{U}^{\mathrm{T}}a)^{\mathrm{T}}\Big]\tilde{U}^{\mathrm{T}} = \tilde{U}[\tilde{D} - \alpha^{-1}vv^{\mathrm{T}}]\tilde{U}^{\mathrm{T}} \tag{3.8}$$

and

$$\alpha = r + \sum_{i=1}^{n} \tilde{d}_i f_i^2 = r + \sum_{i=1}^{n} f_i v_i \tag{3.9}$$

Incidentally, note that the components of v are simply

$$v_i = \tilde{d}_i f_i, \qquad i = 1, \ldots, n \tag{3.10}$$

Now observe that if the bracketed term of (3.8) is factored as $\overline{U}\overline{D}\overline{U}^{\mathrm{T}}$, then (since the product of unit upper triangular matrices is again unit upper triangular) it follows that

$$\hat{U} = \tilde{U}\overline{U} \qquad \text{and} \qquad \hat{D} = \overline{D} \tag{3.11}$$

Thus the construction of the updated U–D factors depends on the simpler factorization

$$\overline{U}\hat{D}\overline{U} = \tilde{D} - \alpha^{-1} v v^{\mathrm{T}} \tag{3.12}$$

The Agee–Turner factorization update theorem (cf. Chapter III) applied to this problem results in the following rather simple representation for $\overline{U}$ and $\hat{D}$:

$$\hat{d}_j = \tilde{d}_j + c_j v_j^2 \tag{3.13}$$

$$c_{j-1} = c_j \tilde{d}_j / \hat{d}_j \tag{3.14}$$

$$\overline{U}_{ij} = c_j v_j v_i / \hat{d}_j, \qquad i = 1, \ldots, j-1 \tag{3.15}$$

Equations (3.13)–(3.15) are backwards recursive, for $j = n$ down to $j = 1$ and

$$c_n = -1/\alpha$$

Remark: Since all the d's are theoretically positive it follows from (3.14) that all the c_j will have the same sign, and because c_n is negative they will all be negative. The result of this observation is that the updated diagonals $\hat{d}_j$ are computed as differences. Such calculations are susceptible to a loss of accuracy due to cancellation and this might even result in negative $\hat{d}$ terms being computed. For this reason we seek a different representation for $\hat{d}_j$, i.e., Eq. (3.4).

To obtain the desired result (3.4), we rearrange (3.13) and (3.14) as

$$\hat{d}_j = \tilde{d}_j + c_j\left(\tilde{d}_j f_j\right)^2 = \tilde{d}_j\left(1 + c_j \tilde{d}_j f_j^2\right) \tag{3.16}$$

$$1/c_{j-1} = 1/c_j\left(\hat{d}_j/\tilde{d}_j\right) = 1/c_j + \tilde{d}_j f_j^2 \tag{3.17}$$

But since $1/c_n = -\alpha = -(r + \sum_{k=1}^{n} \tilde{d}_i f_i^2)$, it follows that

$$1/c_j = -\alpha_j \tag{3.18}$$

and when this result is substituted into (3.15), the result (3.4) follows.

Turning our attention to the construction of $\hat{U}$, we recall from (3.11) that $\hat{U} = \tilde{U}\overline{U}$, and note that $\overline{U}$ can be written as

$$\overline{U} = I + \left[0 \; \lambda_2 v^{(1)} \; \lambda_3 v^{(2)} \; \cdots \; \lambda_n v^{(n-1)}\right] \tag{3.19}$$

where

$$\lambda_j = -v_j / \left(\alpha_j \hat{d}_j\right)$$

and

$$\left(v^{(j)}\right)^{\mathrm{T}} = (v_1 \cdots v_j \; 0 \cdots 0) \tag{3.20}$$

From (3.1) and (3.4) it follows that λ_j can also be represented as in (3.5).

Gill *et al.* (1972) call triangular matrices of form (3.19) "special" and point out, as we shall show, that matrix products involving such matrices are easily and efficiently evaluated. A direct computation of $\tilde{U}\overline{U}$ using representation (3.19) gives

$$\hat{U} = \tilde{U} + \left[0 \; \lambda_2 \tilde{U} v^{(1)} \; \lambda_3 \tilde{U} v^{(2)} \; \cdots \; \lambda_n \tilde{U} v^{(n-1)}\right] \tag{3.21}$$

The construction claimed in the theorem follows by appropriately labeling the columns of $\hat{U}$ and $\tilde{U}$ and exhibiting the recursion inherent in $K_j = \tilde{U} v^{(j)}$, which can be obtained by exploiting the special structure of $v^{(j)}$, i.e.,

$$K_{j+1} = \tilde{U} v^{(j+1)} = K_j + v_j \tilde{u}_j \tag{3.22}$$

The significance of our $\hat{U}$ computation is twofold; the need for storing the intermediate matrix $\overline{U}$ is circumvented, and computation is reduced by a factor of n (viz., computation of $\overline{U}$, as well as our computation of $\hat{U}$, requires $0.5n(n-1)$ additions and multiplications, and direct computation of $\tilde{U}\hat{U}$ requires $0.5n(n-1)(n-2)$ such operations).

The last assertion of the theorem, Eq. (3.7), is obtained from

$$\begin{aligned} K &= \tilde{U}\tilde{D}\tilde{U}^{\mathrm{T}} a/\alpha && \text{(cf. Eq. (II.6.12))} \\ &= \tilde{U} v/\alpha_n \\ &= K_{n+1}/\alpha_n && \text{(cf. Eq. (3.22))} \end{aligned}$$

Carlson's upper triangular square root factorization is an easy consequence of our result; and we state it as a corollary.

Corollary V.3.1 (Square Root Covariance Update Algorithm) Let $\tilde{P} = \tilde{S}\tilde{S}^{\mathrm{T}}$, where $\tilde{S}$ is upper triangular. An updated upper triangular covariance square root factor $\hat{S}$, and the Kalman gain K can be obtained from the following algorithm:

$$\bar{f} = \tilde{S}^{\mathrm{T}}a, \quad \bar{f}^{\mathrm{T}} = (\bar{f}_1 \cdots \bar{f}_n) \tag{3.23}$$

$$\alpha_0 = r, \qquad K_2^{\mathrm{T}} = (\tilde{S}_{11}\bar{f}_1 \overbrace{0 \cdots 0}^{n-1}) \tag{3.24}$$

For $j = 1, \ldots, n$ recursively cycle through the ordered equations (3.25)–(3.29):

$$\alpha_j = \alpha_{j-1} + \bar{f}_j^2 \tag{3.25}$$

$$\beta_j = (\alpha_{j-1}/\alpha_j)^{1/2}, \qquad \gamma_j = \bar{f}_j/(\beta_j\alpha_j) \tag{3.26}$$

$$\hat{S}_{jj} = \tilde{S}_{jj}\beta_j \tag{3.27}$$

$$\hat{S}_{ij} = \tilde{S}_{ij}\beta_j - \gamma_j K_j(i), \qquad i = 1, \ldots, j-1^{\dagger} \tag{3.28}$$

$$K_{j+1}(i) = K_j(i) + \bar{f}_j\tilde{S}_{ij}, \qquad i = 1, \ldots, j \tag{3.29}$$

and

$$K = K_{n+1}/\alpha_n \tag{3.7}$$

The proof of this result is readily obtained from the theorem by noting that $\hat{S} = \hat{U}\hat{D}^{1/2}$ is an upper triangular square root of P.

Remark: Our arrangement of this algorithm differs from that proposed by Carlson in that we use $(\alpha_{j-1}/\alpha_j)^{1/2}$, with $0 < \alpha_{j-1}/\alpha_j \leqslant 1$, while Carlson uses $(\alpha_{j-1}\alpha_j)^{1/2}$, with $0 < \alpha_{j-1}\alpha_j \leqslant \alpha_j^2$. The α's can be large in problems involving large *a priori* estimate uncertainties, and they can be small in problems involving near perfect measurements and small *a priori* uncertainties. When $\alpha_{j-1}\alpha_j$ is computed in such cases there is a danger of computer overflow and underflow.

FORTRAN mechanizations for both the *U–D* and the triangular square root covariance algorithms are included in Appendix V.A.

†When $j = 1$, Eq. (3.28) is omitted.

V.4 Sequential Data Processing Algorithm Computation Counts and Comparisons

In this section we compare computation counts for the covariance related and information related data processing algorithms, i.e., the Kalman, Potter, U–D, and triangular square root algorithms and the recursive normal equation and SRIF algorithms. It is important to note that the computational requirements are dependent on the algorithm mechanization so that if one uses a different equation mechanization different conclusions will result. An infamous illustration of this statement is the Kalman filter mechanization. Let K and a be n vectors and consider the following equivalent mechanizations of the Kalman covariance update equation $P = (I - Ka^{\mathrm{T}})P(I - Ka^{\mathrm{T}})^{\mathrm{T}} + KK^{\mathrm{T}}$:

$$W_1 = I - Ka^{\mathrm{T}}, \qquad W_1(n, n) \tag{4.1}$$

$$W_2 = W_1 P, \qquad W_2(n, n) \tag{4.2}$$

$$\hat{P} = W_2 W_1^{\mathrm{T}} + KK^{\mathrm{T}} \tag{4.3}$$

or, equivalently,

$$v_1 = Pa, \qquad (v_1 \text{ an } n \text{ vector}) \tag{4.4}$$

$$P_1 = P - Kv_1^{\mathrm{T}}, \qquad P_1(n, n) \tag{4.5}$$

$$v_2 = P_1 a, \qquad (v_2 \text{ an } n \text{ vector}) \tag{4.6}$$

$$\hat{P} = (P_1 - v_2 K^{\mathrm{T}}) + KK^{\mathrm{T}} \tag{4.7}^{\dagger}$$

Both sets (4.1)–(4.3) and (4.4)–(4.7) give the same result, and both can save computation in the final step, (4.3) and (4.7), respectively, by exploiting the symmetry of $\hat{P}$. The second method, however, requires an order of magnitude less computation than the first. The first method requires $1.5n^3 + 2n^2 + n$ multiplications, while the second method requires only $4n^2 + n$ multiplications.

Remark: The first mechanization is frequently used, probably because estimation practioners have often been unaware of the difference in computational requirements. Another reason for its use may be that this

† We do not recommend further simplification of Eq. (4.7) by forming $(K - v_2)K^{\mathrm{T}}$, because numerical experience indicates that when $K \approx v_2$, i.e., when K is the optimal Kalman gain, serious errors can result. See also the FORTRAN mechanization given in Appendix II.D.

mechanization can be used in conjunction with a matrix multiplication computer subroutine and thus *appears* easier to implement.

To avoid operation count ambiguities we tabulated, in Appendix V.B, operation counts corresponding to our suggested FORTRAN mechanizations. Characteristics of particular problems may require additional complexities. For example, in numerically ill-conditioned problems where accuracy and numeric stability are of prime concern, one often normalizes variables and scales observation coefficients. In the SRIF case row and column permutations are employed to enhance accuracy (cf. Lawson and Hanson [2]) and in the Kalman filter case, diagonal bounding and correlation testing are sometimes employed in an effort to preserve positivity of the computed covariance matrix. The analyst must, of course, augment our operation counts by whatever auxiliary operations he includes.

There are several different computational comparisons that can be made, and we restrict our attention to the following two important cases:

(a) The objective of the problem is to produce parameter estimates after processing m measurements.

(b) The problem requires parameter estimates and parameter error variances after m measurements.

Remark: For problems of high dimension one may be interested in only a certain subset of parameters, and may want estimates, variances, and/or covariances corresponding to this subset of parameters. We leave it to the reader to modify appropriately the Potter, U–D, triangular covariance square root, information (normal equations), and SRIF formulations to accommodate these other important situations.

Arithmetic operation counts are summarized in Tables 1 and 2, respectively, which summarize the tabulations of Appendix V.B. Computations related to the measurement terms z and a are not included because they are common to all the algorithms and because they are problem dependent. It is assumed that the information related algorithms, entries 6–8, scale the measurements (an extra nm multiplications are included in the tables) while the covariance algorithms do not normalize the measurements (cf. the footnote on page 77).

Although Tables 1 and 2 are the basis of our comparisons they do not include means for comparing computational times for the various algorithms. Carlson suggests lumping all arithmetic operations together and then comparing algorithms in terms of their sum total of operation counts. This strategy for comparison is unrealistic, since, for example, square roots can take more than 40 times as long as addition. We prefer

Table 1
Operation Counts for Estimate Only after m Observations

Algorithm	Number of additions	Number of multiplications	Number of divisions	Number of square roots
(1) Conventional Kalman $P - Ka^{\mathrm{T}}P$	$(1.5n^2+3.5n)m$	$(1.5n^2+4.5n)m$	m	0
(2) U–D Covariance Factorization, $P = UDU^{\mathrm{T}}$	$(1.5n^2+1.5n)m$	$(1.5n^2+5.5n)m$	nm	0
(3) Triangular Covariance Square Root	$(1.5n^2+3.5n)m$	$(2n^2+5n)m$	$2nm$	nm
(4) Potter Covariance Square Root	$(3n^2+3n)m$	$(3n^2+4n)m$	$2m$	m
(5) Kalman Stabilized $(I - Ka^{\mathrm{T}})P(I - Ka^{\mathrm{T}})^{\mathrm{T}} + KrK^{\mathrm{T}}$	$(4.5n^2+5.5n)m$	$(4n^2+7.5n)m$	m	0
(6) **SRIF, *R* Triangular**	$1.5n^2+2.5n + (n^2+2n)m$	$2n^2+3n + (n^2+3n)m$	$2n$	n
(7) Normal Equations	$(n^3+6n^2+5n)/6 + (0.5n^2+1.5n)m$	$(n^3+9n^2-4n)/6 + (0.5n^2+2.5n)m$	$2n$	n
(8) **SRIF[a], *R* General**	$(4n^3+3n^2+5n)/6 + n^2m$	$(4n^3+9n^2-n)/6 + n^2m$	$2n$	n

[a] cf. Kaminski *et al.* [3].

instead to assign relative weights to the arithmetic operations and use as a model the UNIVAC 1108 single precision operation times:

$$\tau_{+} = 1.875 \quad \mu\text{sec}, \qquad \tau_{\div} = 8.375 \quad \mu\text{sec},$$

$$\tau_{\times} = 2.625 \quad \mu\text{sec}, \qquad \tau_{\surd} = 40.1 \quad \mu\text{sec}$$

Thus we take

$$\tau_{\times} = 1.4\tau_{+}, \qquad \tau_{\div} = 4.5\tau_{+}, \qquad \tau_{\surd} = 21.4\tau_{+} \tag{4.8}$$

These relative weightings are used to consolidate the columns of Tables 1 and 2. The results, Tables 3 and 4 and 5 and 6 which are derived from

Table 2
Operation Counts for Estimates and Variances after m Observations

Algorithm	Number of additions	Number of multiplications	Number of divisions	Number of square roots
(1) Conventional Kalman $P - Ka^{\mathrm{T}}P$	$(1.5n^2+3.5n)m$	$(1.5n^2+4.5n)m$	m	0
(2) U–D Covariance Factorization $P = UDU^{\mathrm{T}}$	$0.5n^2-0.5n$ $+(1.5n^2+1.5n)m$	n^2-n $+(1.5n^2+5.5n)m$	nm	0
(3) Triangular Covariance Square Root	$0.5n^2+0.5n$ $+(1.5n^2+3.5n)m$	$0.5n^2+0.5n$ $+(2n^2+5n)m$	$2nm$	nm
(4) Potter Covariance Square Root	$n^2+(3n^2+3n)m$	$n^2+(3n^2+4n)m$	$2m$	m
(5) Kalman Stabilized $(I - Ka^{\mathrm{T}})P(I - Ka^{\mathrm{T}})^{\mathrm{T}}$ $+ KrK^{\mathrm{T}}$	$(4.5n^2+5.5n)m$	$(4n^2+7.5n)m$	m	0
(6) SRIF, R Triangular	$(n^3+9n^2+26n)/6$ $+(n^2+2n)m$	$(n^3+15n^2+20n)/6$ $+(n^2+3n)m$	$2n$	n
(7) Normal Equations	$(2n^3+3n^2+7n)/6$ $+(0.5n^2+1.5n)m$	$(2n^3+12n^2-2n)/6$ $+(0.5n^2+2.5n)m$	$2n$	n
(8) SRIF, R General	$(5n^3+3n^2+17n)/6$ $+n^2m$	$(5n^3+15n^2+4n)/3$ $+n^2m$	$2n$	n

them, are thus oriented toward the UNIVAC 1108 computer. The reader involved with a computer with different relative characteristics can easily modify our tabulations to suit his problem. (It will be pointed out shortly, however, that our conclusions depend mostly on $\tau_{\times}$, and that τ_{+} and $\tau_{\sqrt{}}$ generally influence the results only for moderately small n.)

Remark: The decision not to parametrize the comparisons in terms of τ_{+}, $\tau_{\times}$, τ_{+}, and $\tau_{\sqrt{}}$ was the complexity of the ensuing results. By assuming nominal values for these parameters we were able to construct Tables 5 and 6, which make relative comparisons transparent.

Other factors that influence computer timing (such as the operations involved with computer logic, indexing of variables, data transferals, and

Table 3

Weighted Operation Counts for Estimates Only after m Observations

Algorithm	Weighted operation count $* \tau_+$
(1) Conventional Kalman $P - Ka^{\mathrm{T}}P$	$(3.6n^2 + 9.8n + 4.5)m$
(2) U–D Covariance Factorization	$(3.6n^2 + 13.7n)m$
(3) Triangular Covariance Square Root	$(4.3n^2 + 40.9n)m$
(4) Potter Square Root	$(7.2n^2 + 8.6n + 30.4)m$
(5) Kalman Stabilized $(I - Ka^{\mathrm{T}})P(I - Ka^{\mathrm{T}})^{\mathrm{T}} + KrK^{\mathrm{T}}$	$(10.1n^2 + 16.0n + 4.5)m$
(6) SRIF, R Triangular	$4.3n^2 + 37.1n + (2.4n^2 + 6.2n)m$
(7) Normal Equations	$0.4n^3 + 3.1n^2 + 30.3n + (1.2n^2 + 5.0n)m$
(8) SRIF, R General	$1.6n^3 + 2.6n^2 + 31.0n + 2.4n^2m$

Table 4

Weighted Operation Counts When Estimates and Variances Are Required after m Observations

Algorithm	Weighted operation count $* \tau_+$
(1) Conventional Kalman $P - Ka^{\mathrm{T}}P$	$(3.6n^2 + 9.8n + 4.5)m$
(2) U–D Factorization	$1.9n^2 - 1.9n + (3.6n^2 + 13.7n)m$
(3) Triangular Covariance Square Root	$1.2n^2 + 1.2n + (4.3n^2 + 40.9n)m$
(4) Potter Square Root	$2.4n^2 + (7.2n^2 + 8.6n + 30.4)m$
(5) Kalman Stabilized $(I - Ka^{\mathrm{T}})P(I - Ka^{\mathrm{T}})^{\mathrm{T}} + KrK^{\mathrm{T}}$	$(10.1n^2 + 16.0n + 4.5)m$
(6) SRIF, R Triangular	$0.4n^3 + 5.0n^2 + 39.4n + (2.4n^2 + 6.2n)m$
(7) Normal Equations	$0.8n^3 + 3.3n^2 + 31.1n + (1.2n^2 + 5.0n)m$
(8) SRIF, R General	$2.0n^3 + 6.0n^2 + 34.0n + 2.4n^2m$

Table 5 Estimate Only after m Observations[a]

Algorithm	$n = 10$	$n = 15$	$n = 20$	$n = 30$	$n = 50$	$n \to \infty$
(1) Conventional Kalman $P - Ka^{T}P$	$462.5m$	$961.5m$	$1640.5m$	$3538.5m$	$9494.5m$	
(2) U–D Covariance Factorization $P = UDU^{T}$	1.07	1.06	1.04	1.03	1.02	1.00
(3) Triangular Covariance Square Root	1.81	1.64	1.55	1.44	1.35	1.19
(4) Potter Covariance Square Root	1.81	1.85	1.88	1.91	1.94	2.00
(5) Kalman Stabilized $(I - Ka^{T})P(I - Ka^{T})^{T} + KrK^{T}$	2.54	2.62	2.66	2.71	2.74	2.81
(6) SRIF R Triangular	$0.65 + 1.73/m$ (<1 if $m > 4$)	$0.66 + 1.58/m$ (<1 if $m > 4$)	$0.66 + 1.50/m$ (<1 if $m > 4$)	$0.66 + 1.41/m$ (<1 if $m > 4$)	$0.66 + 1.33/m$ (<1 if $m > 3$)	$0.667 + 1.19/m$ (<1 if $m > 3$)
(7) Normal Equations	$0.37 + 2.19/m$ (<1 if $m > 3$)	$0.36 + 2.60/m$ (<1 if $m > 4$)	$0.35 + 3.08/m$ (<1 if $m > 4$)	$0.35 + 4.10/m$ (<1 if $m > 6$)	$0.34 + 6.22/m$ (<1 if $m > 9$)	$0.33 + 0.11n/m$ (<1 if $m > 0.16\,n$)
(8) SRIF R General	$0.52 + 4.69/m$ (<1.5 if $m > 4$)	$0.56 + 6.71/m$ (<1.5 if $m > 7$)	$0.59 + 8.81/m$ (<1.5 if $m > 9$)	$0.61 + 13.13/m$ (<1.5 if $m > 14$)	$0.63 + 21.91/m$ (<1.5 if $m > 25$)	$0.66 + 0.44n/m$ (<1.5 if $m > 0.52\,n$)

[a] Kalman normalized operation counts.

Table 6 Estimates and Variances after m Observations[a]

Algorithm	$n = 10$	$n = 15$	$n = 20$	$n = 30$	$n = 50$	$n \to \infty$
(1) Conventional Kalman $P - Ka^{\mathrm{T}}P$	$462.5m$	$961.5m$	$1640.5m$	$3538.5m$	$9494.5m$	
(2) U–D Covariance Factorization $P = UDU^{\mathrm{T}}$	$1.07+0.37/m$	$1.06+0.41/m$	$1.04+0.44/m$	$1.03+0.47/m$	$1.02+0.49/m$	$1.00+0.53/m$
(3) Triangular Covariance Square Root	$1.81+0.29/m$	$1.64+0.30/m$	$1.55+0.31/m$	$1.44+0.32/m$	$1.35+0.32/m$	$1.19+0.33/m$
(4) Potter Covariance Square Root	$1.81+0.52/m$	$1.85+0.56/m$	$1.88+0.59/m$	$1.91+0.61/m$	$1.94+0.63/m$	$2.00+0.66/m$
(5) Kalman Stabilized $(I - Ka^{\mathrm{T}})P(I - Ka^{\mathrm{T}})^{\mathrm{T}} + KrK^{\mathrm{T}}$	2.54	2.62	2.66	2.71	2.74	2.81
(6) SRIF R Triangular	$0.65+4.09/m$ (<1 if $m > 11$)	$0.66+3.12/m$ (<1 if $m > 9$)	$0.66+3.65/m$ (<1 if $m > 10$)	$0.66+4.66/m$ (<1 if $m > 13$)	$0.66 + 6.79/m$ (<1 if $m > 19$)	$0.667 + 0.11n/m$ (<1 if $m > 0.3\,n$)
(7) Normal Equations	$0.37+3.12/m$ (<1 if $m > 5$)	$0.36+4.07/m$ (<1 if $m > 6$)	$0.35+5.09/m$ (<1 if $m > 8$)	$0.35+7.21/m$ (<1 if $m > 11$)	$0.34+11.57/m$ (<1 if $m > 17$)	$0.33+0.22n/m$ (<1 if $m > 0.3\,n$)
(8) SRIF R General	$0.52+6.36/m$ (<1.5 if $m > 6$)	$0.56+8.95/m$ (<1.5 if $m > 9$)	$0.59+11.63/m$ (<1.5 if $m > 12$)	$0.61+17.08/m$ (<1.5 if $m > 19$)	$0.63+28.09/m$ (<1.5 if $m > 32$)	$0.66+0.56n/m$ (<1.5 if $m > 0.67\,n$)

[a] Kalman normalized operation counts.

input/output), while certainly important, were omitted. Our reasons for not including items of this nature is that most of these expenditures are common† to all of the algorithms and moreover such timing expenditures are not only machine dependent, they are installation dependent, and thus are extremely difficult to quantify.

We normalized the operation counts in Tables 5 and 6 (i.e., the entries in rows 3–8 are ratios of the corresponding operation times divided by the Kalman optimal operation times) because the main purpose of the algorithm tabulations is for comparison. Choosing the conventional Kalman algorithm $P - Ka^{\mathrm{T}}P$ as a reference‡ facilitates comparisons. Tables 5 and 6 give rise to some interesting observations and conclusions.

The covariance-related and information-related algorithms are separated; and each set is ordered in terms of relative efficiency. These orderings are not dependent on the relative arithmetic weights τ_+ and $\tau_{\sqrt{}}$. The reason for this is that the orderings were determined by examining only the highest powers of n (asymptotic behavior) and these terms only involve the relative arithmetic weight τ_x. Note that for $n = 10$ the triangular square root covariance algorithm which requires n square roots and $2n$ divides per measurement differs from its asymptotic value by 65% (cf. Table 5). This variation points up the significance of accounting for the lower order terms in our arithmetic operation counts. For n small these lower order terms can significantly influence relative computational efficiency.

The U–D computational efficiency (cf. Table 5) is, however, virtually indistinguishable from that of the conventional Kalman (when other common computer costs are included). This statement holds for n both small and large. When variance computations are included (cf. Table 6), the results become m dependent, and relative efficiency is penalized by an additive factor of approximately $(50/m)$%. For $m > 4$ the variance computation is negligible.§

The triangular covariance square root algorithm (cf. Table 5) is except for the case $n = 10$, consistently more efficient than Potter's covariance square root. For n small the relative difference is negligible, but for n large

†The information-related algorithms are more complicated than are their covariance-related counterparts because of the additional code and logic required for estimate and variance computations. These items are not generally significant, except in minicomputer-related applications.

‡Comparisons of the other algorithms with the conventional Kalman are made so that one can assess the relative cost of using the more accurate and stable algorithms. Incidently, we do not mean to suggest that the stabilized Kalman filter (algorithm 5) or the normal equation filter (algorithm 7) have accuracy or stability characteristics akin to the square root algorithms.

§Algorithms having operation counts that agree to within 20% are generally indistinguishable when implemented on a digital computer.

the relative difference approaches 70%. This conclusion remains the same when variance computations are included (cf. Table 6).

It is well known that when m is "sufficiently" large the information-related algorithms are more efficient than are the covariance-related algorithms (even the conventional Kalman algorithm); and Tables 5 and 6 indicate how large m must be for this to happen. For example, when only estimates are needed, the SRIF with R triangular is more efficient that the conventional Kalman algorithm when $m > 4$; and this result holds for all n. The normal equation result is n dependent, which is the general state of affairs, and is more efficient (cf. Table 5) when $m > n/6$ (approx). The information-related algorithms perform markedly more computation when variances are required because they then involve explicit[†] matrix inversion. Even in this case, however, the SRIF with triangular R and the normal equation method are still more efficient than the conventional Kalman algorithm, but now m must be larger than before, i.e., now $m > n/3$ (approx) (cf. Table 6).

The SRIF with general R appears in problems that involve time propagation (mapping) and process noise. When R is general it is relatively expensive to compute estimates and/or variances. As indicated in the tables, however, the computations are within 50% of the Kalman optimal when $m > n/2$, estimate only; and $m > \frac{2}{3}n$ estimates and variances.

Remark: It is important to note than when n is large problems with time propagation often involve many bias parameters. In such cases R is block triangular with a triangular lower block (cf. Chapter VII). SRIF's with this special structure are very nearly triangular.

The main points of this discussion are that the U–D covariance factorization is very nearly as efficient as the conventional Kalman algorithm, estimates for the SRIF with triangular R can be obtained even more efficiently than with the conventional Kalman algorithm, and variance computations while costly are not generally prohibitive.

V.5 Filter Algorithm Numerical Deterioration; Some Examples

Kaminski *et al.* [3] note that the condition number[‡] of a matrix square root is the square root of the condition number of the matrix. It is inferred

[†]When only estimates are required one can solve linear equations of the form $Bx = y$. This requires considerably less arithmetic than first computing B^{-1} and then forming $B^{-1}y$.

[‡]Condition numbers (cf. Lawson and Hanson [2]) are used in numerical analysis to measure the sensitivity of matrix calculations to element perturbations.

from this that square root estimation algorithms have essentially twice the numerical precision of their non-square root counterparts. It is important to note that this conclusion is merely an inference and is, except in certain examples, unproven. Rigorous, general analyses of the relative numerical accuracies of the various algorithms is a difficult problem that has not as yet been resolved. The difficulty and complexity of such a task is illustrated by the lengthy SRIF error analyses that appear in the book by Lawson and Hanson [2]. We will not attempt such analyses in this book, but will instead endeavor to convey an understanding of how numeric deterioration affects the various algorithms by examing two particular, insightful examples.

Numerical difficulties often appear in problems that involve many parameters and lengthy computations. We choose, however, not to use complex or unwieldy problems for our examples and instead illustrate certain of the key numerical difficulties by examing simple two-parameter problems. The effect that numerical errors have on an algorithm depends to a large extent on the details of its computer implementation. Thus to determine how an algorithm deteriorates the relevant steps of the algorithm mechanization must be examined. The effects of roundoff and finite word length are simulated in our examples by using a small positive parameter ϵ. It is assumed that ϵ is large enough so that $1 + \epsilon$, when rounded, is greater than one, but that $1 + \epsilon^2 \overset{r}{=} 1$ ($a \overset{r}{=} b$ means that b is the rounded value of a). For example, in a computer with ten-digit accuracy one could take ϵ such that $10^{-9} < \epsilon < 10^{-5}$.

Example V.1 (Singularity Induced by the Information Filter) This example due to Golub (1965) consists of the system of equations

$$\begin{bmatrix} z_1 \\ z_2 \\ z_3 \end{bmatrix} = \begin{bmatrix} 1 & 1-\epsilon \\ 1-\epsilon & 1 \\ 1 & 1 \end{bmatrix} \begin{bmatrix} x_1 \\ x_2 \end{bmatrix} + \begin{bmatrix} \nu_1 \\ \nu_2 \\ \nu_3 \end{bmatrix} \tag{5.1}$$

where the ν vector is normalized.

The evident difficulty with this problem is that the three measurements are nearly linearly dependent. Systems of this type with near linear dependence are sometimes said to have poor observability; and appear, for example, in problems where closely spaced range measurements are used to determine the position of a vehicle.

Although (5.1) has a well-defined least squares solution the problem degenerates when the normal matrix is formed:

$$\Lambda = A^{\mathrm{T}}A = \begin{bmatrix} 2 + (1-\epsilon)^2 & 3 - 2\epsilon \\ 3 - 2\epsilon & 2 + (1-\epsilon)^2 \end{bmatrix}$$

$$\stackrel{r}{=} \begin{bmatrix} 3 - 2\epsilon & 3 - 2\epsilon \\ 3 - 2\epsilon & 3 - 2\epsilon \end{bmatrix} \tag{5.2}$$

The rounded normal matrix is obviously singular so that the estimate $\Lambda^{-1}A^{\mathrm{T}}z$ cannot be computed. Evidently the information filter algorithm induced numerical degeneracy into the problem.

An Aside: Although information has been destroyed by the formation of (5.2), one could still seek an approximate least squares solution by introducing the pseudoinverse of Λ (cf. Noble (1969) or Lawson and Hanson [2]). It is not our intention in this book to discuss pseudoinverses, but nevertheless we mention, for reference purposes, that the pseudoinverse Λ^{+} for this example is

$$\Lambda^{+} = \frac{1}{4(3-2\epsilon)} \begin{bmatrix} 1 & 1 \\ 1 & 1 \end{bmatrix} \tag{5.3}$$

and the corresponding estimates for x_1 and x_2 are

$$x_1^{+} = x_2^{+} = \frac{1}{3-2\epsilon}\left(\left(1 - \frac{\epsilon}{2}\right)\frac{(z_1 + z_2)}{2} + \frac{z_3}{2}\right) \tag{5.4}$$

the superscript + is used to accentuate the fact that these estimates were obtained using the pseudoinverse. Estimates (5.4) appear to be an acceptable approximate solution, but they are, in fact, quite different from the actual least squares estimates $\hat{x}_1$ and $\hat{x}_2$:

$$\begin{aligned} \hat{x}_1 &= \frac{1}{3 - 2\epsilon + \epsilon^2/2}\left(\frac{(3-\epsilon)z_1 - [3(1-\epsilon) + \epsilon^2]z_2}{2\epsilon} + \frac{z_3}{2}\right) \\ \hat{x}_2 &= \frac{1}{3 - 2\epsilon + \epsilon^2/2}\left(\frac{(3-\epsilon)z_2 - [3(1-\epsilon) + \epsilon^2]z_1}{2\epsilon} + \frac{z_3}{2}\right) \end{aligned} \tag{5.5}$$

By comparing (5.4) and (5.5) it becomes clear that the errors introduced by the information algorithm are serious and cannot be overcome by introducing the pseudoinverse. ■

Suppose now that (5.1) is transformed to a triangular form using orthogonal transformations, i.e., if

$$[A \ \ z] = \begin{bmatrix} 1 & 1-\epsilon & z_1 \\ 1-\epsilon & 1 & z_2 \\ 1 & 1 & z_3 \end{bmatrix}$$

then

$$T[A \ \ z] = \begin{bmatrix} R(1,1) & R(1,2) & \hat{z}_1 \\ 0 & R(2,2) & \hat{z}_2 \\ 0 & 0 & e \end{bmatrix} \tag{5.6}$$

where $TT^{\mathrm{T}} = I$.

We use this example to show how one can compute R and z directly from the array $[A \ \ z]$, without using the Householder transformations of Chapter IV. This method, communicated to the author by Dr. H. Lass, is quite amenable to hand computation. Lass's method is based on the observation that inner products are preserved by orthogonal transformations, i.e., if $x_1 = Tx$ and $y_1 = Ty$, then $x_1^{\mathrm{T}} y_1 = x^{\mathrm{T}} y$. By systematically applying this idea one can compute $[R \ \ z]$ a row at a time. To illustrate the method we successively compute inner products of the first column with itself, then with the second column, and finally with the last column. The inner products are, of course, computed using both the $[R \ \ z]$ and $[A \ \ z]$ arrays:

$$R(1,1)^2 = 2 + (1-\epsilon)^2$$

$$R(1,1)R(1,2) = 3 - 2\epsilon$$

$$R(1,1)\hat{z}_1 = z_1 + (1-\epsilon)z_2 + z_3.$$

Thus setting $p = (2 + (1-\epsilon)^2)^{1/2}$ we have the elements of row 1:

$$\begin{aligned} R(1,1) &= p \\ R(1,2) &= (3-2\epsilon)/p \\ \hat{z}_1 &= (z_1 + (1-\epsilon)z_2 + z_3)/p \end{aligned} \tag{5.7}$$

Next, the inner products of column two with itself and then with column three are computed:

$$R(1,2)^2 + R(2,2)^2 = 2 + (1-\epsilon)^2$$

$$R(1,2)\hat{z}_1 + R(2,2)\hat{z}_2 = (1-\epsilon)z_1 + z_2 + z_3$$

Thus we have the elements of row 2:

$$\begin{aligned} R(2,2) &= \epsilon(6 - 4\epsilon + \epsilon^2)^{1/2}/p \\ \hat{z}_2 &= [(1-\epsilon)z_1 + z_2 + z_3 - R(1,2)\hat{z}_1]/R(2,2) \end{aligned} \tag{5.8}$$

Finally, the residual of the process, e, can be obtained by taking the inner product of the last column with itself:

$$\hat{z}_1^2 + \hat{z}_2^2 + e^2 = z_1^2 + z_2^2 + z_3^2$$

i.e.,

$$e^2 = z_1^2 + z_2^2 + z_3^2 - \hat{z}_1^2 + \hat{z}_2^2$$

Unfortunately this method of constructing R is tantamount to applying a Cholesky factorization (cf. Chapter III) to the normal matrix. Thus this method, although apparently of square root type, has the same numeric problem as the information filter algorithm. If this algorithm were computer mechanized the effects of numeric errors would result in the appearance of a computed $R(2,2)$ of zero.

Let us now apply Golub's triangularization method using Householder transformations (cf. Theorem IV.3.1) and see how computation of a singular R matrix is avoided. As a prelude to studying the construction we remind the reader that triangular factorization is unique (except for the signs chosen for the square roots) and thus when error free computations are performed the Golub and Lass results coincide.

The first Householder transformation T_1, chosen to null the entries of the first column of A, results in

$$T_1 \begin{bmatrix} 1 & 1-\epsilon & z_1 \\ 1-\epsilon & 1 & z_2 \\ 1 & 1 & z_3 \end{bmatrix} = \begin{bmatrix} R(1,1) & R(1,2) & \hat{z}_1 \\ 0 & A_2(2,2) & A_2(2,3) \\ 0 & A_2(3,2) & A_2(3,3) \end{bmatrix} \tag{5.9}$$

where

$$R(1, 1) = -p = -\left(2 + (1-\epsilon)^2\right)^{1/2}$$

$$R(1, 2) = (1-\epsilon) - [(1-\epsilon)(p+2) + 1]/p$$

$$\begin{bmatrix} A_2(2, 2) \\ A_2(3, 2) \end{bmatrix} = \begin{bmatrix} 1 \\ 1 \end{bmatrix} - \frac{(1-\epsilon)(p+2)+1}{p(p+1)} \begin{bmatrix} 1-\epsilon \\ 1 \end{bmatrix} \tag{5.10}$$

$$\lambda = (p+1)z_1 + (1-\epsilon)z_2 + z_3$$

$$\hat{z}_1 = z_1 - \lambda/p$$

$$\begin{bmatrix} A_2(2, 3) \\ A_2(3, 3) \end{bmatrix} = \begin{bmatrix} z_2 \\ z_3 \end{bmatrix} - \frac{\lambda}{p(p+1)} \begin{bmatrix} 1-\epsilon \\ 1 \end{bmatrix}$$

These results are a direct application of the formulae of Appendix IV.A. Terms have not been combined because we want the reader to be able to trace the arithmetic operations as they would be executed in a computer. If we denote computed quantities with the symbol "‾", we have

$$p \overset{r}{=} (3 - 2\epsilon)^{1/2} = \bar{p} \tag{5.11}$$

$$\bar{R}(1, 1) = -\bar{p}, \qquad \bar{R}(1, 2) = -\bar{p}$$
$$\bar{\lambda} = (\bar{p}+1)z_1 + (1-\epsilon)z_2 + z_3, \qquad \bar{z} = z_1 - \bar{\lambda}/\bar{p} \tag{5.12}$$

$$\begin{bmatrix} \bar{A}_2(2, 2) & \bar{A}_2(2, 3) \\ \bar{A}_2(3, 2) & \bar{A}_2(3, 3) \end{bmatrix} = \begin{bmatrix} \epsilon(\bar{p}+1)/2^* & z_2 - \bar{\lambda}(1-\epsilon)/[\bar{p}(\bar{p}+1)] \\ \epsilon/(\bar{p}+1)^* & z_3 - \bar{\lambda}/[\bar{p}(\bar{p}+1)] \end{bmatrix} \tag{5.13}$$

The form of Eq. (5.10) indicates that the $*$ terms suffer somewhat from cancellation due to finite computer word length, i.e., they are rounded to a lesser number of digits than the machine word length. Our assumption on the size of ϵ, however, prevents these terms from being completely nulled and thus we are assured that $\bar{R}(2, 2)$ will be nonzero when it is computed.

Remark: To amplify on this allegation we note that if the computer word length were 10 decimal digits and $\epsilon = 10^{-6}$, then the computer representation of $1 - \epsilon(\bar{p} + 1)/2$ would contain only four digits of the number $\epsilon(\bar{p} + 1)/2$. Since $\bar{A}_2(2, 2) = 1 - (1 - \epsilon(\bar{p} + 1)/2)$ it follows that $\bar{A}_2(2, 2)$ would be accurate to only four digits.

Most modern computers are provided with an option to accumulate inner products using extra† precision. If that were done in this example, there would be little or no error in the $\bar{A}$ computation due to cancellation. Including extra precision accumulation is recommended for all algorithms because it adds negligibly to computer execution time and has no effect on storage requirements. Incidentally we note that inclusion of extra precision accumulation would not have helped the information filter algorithm or Lass's method; the reason being that these algorithms failed because of roundoff and not cancellation.

The high points of this example are that:

(a) It illustrates how algorithm numerics can turn a near singular problem into a singular one. It was also shown that numerically induced singularities can not be circumvented by introducing the pseudoinverse.

(b) Lass's method for constructing a triangular factorization was shown by example to be useful for hand computations; his method gave $[R \;\; \hat{z}]$ (cf. Eqs. (5.7) and (5.8)).

(c) Golub's triangularization method, which is fundamental to the SRIF, was demonstrated.

To better understand the numerics of this example the reader is advised to complete Golub's SRIF solution on a hand calculator using an appropriate ϵ, and then to compare the results with R and $\hat{z}$ numerically computed from Eqs. (5.7) and (5.8). Most importantly, the computed estimates $R^{-1}\hat{z}$ should be compared with the solution, Eqs. (5.5).

Example V.2 (Kalman Algorithm Deterioration Due to Large *a Priori* Statistics and Poor Observability) Suppose that the initial estimates of x_1 and x_2 are zero and that $P = \sigma^2 I$ where $\sigma = 1/\epsilon$ is large. Using a large *a priori* variance is a common ploy that expresses an initial lack of knowledge about the system. Suppose now that the following two observations are to be processed:

$$\begin{bmatrix} z_1 \\ z_2 \end{bmatrix} = \begin{bmatrix} 1 & \epsilon \\ 1 & 1 \end{bmatrix} \begin{bmatrix} x_1 \\ x_2 \end{bmatrix} + \begin{bmatrix} \nu_1 \\ \nu_2 \end{bmatrix} \tag{5.14}$$

where the ν's are normalized.

†"Extra" in this context generally means using an accumulator having double word length.

The four principal covariance algorithms are used to solve this problem and the results, tabulated in Table 7, merit discussion.

Remark: The triangular covariance square root and the information-related algorithms were not used to solve this example because the former algorithm has the same numerical character as the $U\text{–}D$ covariance factorization (which we do discuss) and the latter algorithms are known to fare well on problems of this nature.

The conventional Kalman algorithm computed covariance has completely degenerated and if more observations were processed the computed variances could become increasingly more negative. Examples of this nature point up the instability inherent in this algorithm. Estimation practitioners who use this algorithm generally try to impede covariance degeneration by modifying the algorithm so that updated diagonal elements cannot be too small. Off-diagonal covariance matrix elements are also at times a source of error, and these are bounded by limiting the size of correlations. Selecting appropriate tolerance parameters is problem dependent and generally ad hoc.

The stabilized Kalman algorithm gives good results for the computed covariance and this illustrates why it is recommended over the conventional Kalman algorithm. The $U\text{–}D$ factorization performed well and gave computed results that agree well with the exact values. Potter's algorithm appeared to do well also, although the unwieldly form of the exact S_2 covariance factor prevented a quantitative comparison of terms. Evidently the $U\text{–}D$ and Potter algorithms gave comparable gain and covariance results, with the $U\text{–}D$ being a trifle closer to the exact values.

An Aside: Thornton and Bierman [6], studied numerical deterioration of the Kalman algorithm in considerable detail. Using nontrivial examples they have shown, among other things, that (a) even the stabilized Kalman mechanizations (cf. Appendix II.D) can give negative diagonal computed results; (b) "patching" techniques for the conventional Kalman algorithm (bounding of diagonal and off-diagonal elements) gives only spotty improvement and sometimes even worsens the results; and (c) in cases that one would not expect, computational errors are a dominant error and can be larger than the errors due to mismodeling. The Potter and $U\text{–}D$ algorithms performed better than the Kalman covariance algorithms in every case of their study and the results were such that numerical errors introduced using the factorization algorithms had a negligible effect on system performance. ■

Incidentally, we note that one of the characteristics of triangular factorization algorithms is that singularities and near singularities are generally easily detected by monitoring the diagonal factor elements. In the

Table 7

Numeric Comparisons for Example 2

Filter parameters	Exact values	Conventional Kalman
K_1 Gain	$\frac{1}{\alpha}\begin{bmatrix} 1 \\ \epsilon \end{bmatrix}, \alpha = 1 + 2\epsilon^2$	$\begin{bmatrix} 1 \\ \epsilon \end{bmatrix}$
P_1 Covariance Factors	$D_1 = 1/(1+\epsilon^2)$, $D_2 = \sigma^2(1+\epsilon^2)/\alpha$ $U(1,2) = -\epsilon/(1+\epsilon^2)$	—
	$S_1 = \frac{1}{1+\epsilon^2}\begin{bmatrix} \epsilon + 1/\sqrt{\alpha} & -(1 - \epsilon/\sqrt{\alpha}) \\ -(1 - \epsilon/\sqrt{\alpha}) & \epsilon^2/\sqrt{\alpha} \end{bmatrix}$	—
P_1 Covariance	$P_1 = \frac{1}{\alpha}\begin{bmatrix} 2 & -\sigma \\ -\sigma & \sigma^2 + 1 \end{bmatrix}$	$\begin{bmatrix} 0 & -\sigma \\ \text{—}^{b} & \sigma^2 \end{bmatrix}$
K_2 Gain	$K_2 = \frac{1}{\Delta}\begin{bmatrix} -\epsilon(1 - 2\epsilon) \\ 1 - \epsilon + \epsilon^2 \end{bmatrix}$ $\Delta = 1 - 2\epsilon + 2\epsilon^2(2+\epsilon^2)$	$\frac{1}{\Delta_r}\begin{bmatrix} -\epsilon \\ 1 - \epsilon \end{bmatrix}$ $\Delta_r = 1 - 2\epsilon$
P_2 Covariance Factors	$D_1 = 1/(2+\epsilon^2)$, $D_2 = (2+\epsilon^2)/\Delta$ $U(1,2) = -(1+\epsilon)/(2+\epsilon^2)$	—
	—[c]	—
P_2 Covariance	$P_2 = \frac{1}{\Delta}\begin{bmatrix} 1 + 2\epsilon^2 & -(1+\epsilon) \\ -(1+\epsilon) & 2 + \epsilon^2 \end{bmatrix}$	$\frac{1}{\Delta_r}\begin{bmatrix} -1 & +1 \\ \text{—}^{b} & 0 \end{bmatrix}$

[a] These covariances are not part of the algorithm. They are included for comparison purposes only.

[b] Only the upper triangular covariance elements are computed.

[c] The exact representation of S_2 is too unwieldy to be included.

$$\left(P_0 = \sigma^2 I,\ z = \begin{pmatrix} 1 & \epsilon \\ 1 & 1 \end{pmatrix} x + v\right)$$

U–D Covariance factorization	Potter covariance square root	Stabilized Kalman
$\begin{bmatrix} 1 \\ \epsilon \end{bmatrix}$	$\begin{bmatrix} 1 \\ \epsilon \end{bmatrix}$	$\begin{bmatrix} 1 \\ \epsilon \end{bmatrix}$
$D_1 = 1,\ D_2 = \sigma^2$ $U(1, 2) = -\epsilon$	—	—
—	$S_1 = \dfrac{1}{1+\epsilon}\begin{bmatrix} 1 & -1 \\ -1 & \sigma + 1 \end{bmatrix}$	—
$\begin{bmatrix} 2 & -\sigma \\ —^{b} & \sigma^2 \end{bmatrix}^{a}$	$\begin{bmatrix} 2/(1+2\epsilon) & -\sigma \\ —^{b} & \sigma^2 \end{bmatrix}^{a}$	$\begin{bmatrix} 2 & -\sigma \\ —^{b} & \sigma^2 \end{bmatrix}$
$\begin{bmatrix} -\epsilon \\ (1-\epsilon)/\Delta_r \end{bmatrix}$	$\begin{bmatrix} -\epsilon \\ (1-\epsilon)/\Delta_r \end{bmatrix}$	$\begin{bmatrix} -\epsilon \\ (1-\epsilon)/\Delta_r \end{bmatrix}$
$D_1 = 1/2,\ D_2 = 2/\Delta_r$ $U(1, 2) = -(1+\epsilon)/2$	—	—
—	$\bar{S}_2 = \begin{bmatrix} (1+\epsilon)^{-1} & -\epsilon \\ -(1+\epsilon)^{-1} & (1+\epsilon)^{-1} \end{bmatrix}$	—
$\dfrac{1}{\Delta_r}\begin{bmatrix} 1 & -(1+\epsilon) \\ —^{b} & 2 \end{bmatrix}^{a}$	$\begin{bmatrix} (1+2\epsilon)^{-1} & -(1-3\epsilon)/\Delta_r \\ —^{b} & 2 \end{bmatrix}$	$\begin{bmatrix} 1+2\epsilon & -(1+\epsilon)/\Delta_r \\ —^{b} & (2-\epsilon)/\Delta_r \end{bmatrix}$

U–D factorization det $P = D(1) \cdots D(n)$ and thus near singularity requires at least one very small D element. When dealing with higher-dimensional problems near singularity of the covariance matrix is sometimes camouflaged, viz., the innocent-looking three-dimensional example of Section III.2. The U–D factors for the covariance matrix of that example are

$$U = \begin{bmatrix} 1 & \dfrac{0.5 - \delta}{0.5 + \delta} & 0.5 - \delta \\ 0 & 1 & -0.5 + \delta \\ 0 & 0 & 1 \end{bmatrix}$$

$$D = \mathrm{Diag}\left(\delta \, \frac{1.5 - \delta}{0.5 + \delta}, \; 1 - (0.5 - \delta)^2, \; 1 \right)$$

Positivity of the D elements shows that $0 < \delta < 0.5$ and for δ small the near singularity of the covariance is apparent by inspection of d_1. This example points up the significance of monitoring D.

Remark: Estimation practitioners often monitor variance magnitudes to detect near singularity. While small variances are sufficient to cause near singularity they are not a necessary condition. The insidious problem of some parameter subset being near dependent (i.e., nearly unobservable) cannot be determined by perusing variances. Perusal of the D elements (as illustrated above) can, however, expose the existence of near dependent parameter subsets. The singular value decomposition, discussed at length by Lawson and Hanson [2], can then be used to identify which combinations of parameters are nearly unobservable.

APPENDIX V.A U–D and Upper Triangular $P^{1/2}$ FORTRAN Mechanizations

U–D Update

Input x — *A priori* estimate

U — Upper triangular matrix with $D(i)$ stored on the diagonals; this corresponds to the *a priori* covariance

z, a, r — Observation, observation coefficients (n vector), and observation error variance, respectively

Remark: The fact that U theoretically has unit elements on the diagonal is implicit in the mechanization.

Output x Updated estimate

U Updated upper triangular matrix, with the updated $D(i)$ stored on the diagonals; this corresponds to the updated covariance

α The innovations variance of the measurement $z = a^{\mathrm{T}}x + \nu$

b The unweighted Kalman gain $K = b/\alpha$

Remark: To save storage and for brevity, z and a are destroyed in the mechanization.

```
    DO 2 J = 1, n                       @@
2   z = z - a(J) * x(J)                 @@
    DO 10 J = n, 2, - 1
    DO 5 K = 1, J - 1
5   a(J) = a(J) + U(K, J) * a(K)        @ (cf. Eq. (3.1))
10  b(J) = U(J, J) * a(J)               @ (cf. Eq. (3.1))
    b(1) = U(1, 1) * a(1)
```

Comments: $z := z - a^{\mathrm{T}}x$ (the computed residual), $b = DU^{\mathrm{T}}a$, and $a := U^{\mathrm{T}}a$ have all been computed.

```
    α = r + b(1) * a(1)
    γ = 1./α
    U(1, 1) = r * γ * U(1, 1)

    DO 20 J = 2, n
    β = α
    α = α + b(J) * a(J)                 @ (cf. Eq. (3.3))
    λ = -a(J) * γ                       @ (cf. Eq. (3.5))
    γ = 1./α
    U(J, J) = β * γ * U(J, J)           @ (cf. Eq. (3.4))

    DO 20 I = 1, J - 1
    β = U(I, J)
    U(I, J) = β + b(I) * λ              @ (cf. Eq. (3.5))
20  b(I) = b(I) + b(J) * β              @ (cf. Eq. (3.6))
    z = z * γ                           @@
    DO 30 J = 1, n                      @@
30  x(J) = x(J) + b(J) * z              @@
```

Comments: If estimates are not to be computed, then the statements containing "@@" can be omitted. With modest modifications one could use U stored as a vector. This would reduce storage in problems having large dimension.

The steps necessary to update an upper triangular factorization $P = SS^{\mathrm{T}}$ are almost the same as those involved with the U–D update. For reference purposes, however, we also include the FORTRAN mechanization corresponding to this factorization (cf. Eqs. (3.23)–(3.29)).

$S = P^{1/2}$ Covariance Square Root Update

Input x — *A priori* estimate
S — *A priori* upper triangular covariance square root
z, a, r — Observation, observation coefficients, and observation error variance, respectively

Output x — Updated estimate
S — Updated upper triangular covariance square root
b, α — The unweighted Kalman gain, and the innovations residual variance, respectively; $K = b/\alpha$

Comment: To save storage and for brevity of exposition, z and a are destroyed.

```
    DO 10 J = n, 1, - 1
    σ = 0
    z = z - a(J) * x(J)                @@
    DO 5 K = 1, J
 5  σ = σ + S(K, J) * a(K)
10  a(J) = σ                           @ (cf. Eq. (3.23))
```

Comment: $z := z - a^{\mathrm{T}}x$, $a := S^{\mathrm{T}}a$.

```
    α = r + a(1) ** 2
    b(1) = S(1, 1) * a(1)
    S(1, 1) = S(1, 1) * SQRT(r/α)

    DO 20 J = 2, n
    b(J) = S(J, J) * a(J)
    γ = α                              @ γ = α_{j-1}
    α = α + a(J) ** 2                  @ (cf. Eq. (3.25))
    β = SQRT(γ/α)                      @ (cf. Eq. (3.26))
    γ = a(J)/(α * β)                   @ (cf. Eq. (3.26))
    S(J, J) = S(J, J) * β              @ (cf. Eq. (3.27))
    DO 20 K = 1, J - 1
    s = S(K, J)
    S(K, J) = s * β - γ * b(K)         @ (cf. Eq. (3.28))
20  b(K) = b(K) + a(J) * s             @ (cf. Eq. (3.29))
    z = z/α                            @@
    DO 30 J = 1, n                     @@
30  x(J) = x(J) + b(J) * z             @@
```

Note how directly the algorithms translate into FORTRAN code. This is characteristic of the algorithms of this book.

APPENDIX V.B Arithmetic Operation Counts for Various Data Processing Algorithms

In this appendix we tabulate the arithmetic operations that are involved with the covariance-related and information-related data processing algorithms. These operation counts are the basis for the comparisons that are made in Section V.4.

All of the algorithms except the SRIF are assumed to process m measurements of the form

$$z = a^{\mathrm{T}}x + \nu$$

where a is an n vector and ν is zero mean with variance r.

Remark: Operation counts are based on FORTRAN mechanizations. Thus, for example, $\Sigma_{j=1}^{n} a(j)b(j)$ is tabulated as having n additions. The extra addition is due to the FORTRAN mechanization of the inner product.

Estimate and variance operation counts for the SRIF are obtained from Table B.6. The reason for this is that once Λ has been factored both methods use the same technique. A separate table for the SRIF with R nontriangular was not included, although there is a FORTRAN mechanization given in Appendix VII.B. Nontriangular R's come into being when time propagation and process noise are included into the problem; these topics will be discussed in the next chapter.

Table B.1

Kalman Data Processing[a]

$P := P - Ka^{\mathrm{T}}P$

Computation	Remarks	Number of additions	Number of multiplications	Number of divisions
$v = Pa$	v Temporary work vector	n^2	n^2	
$s = 1/(a^{\mathrm{T}}v + r)$	s^{-1} Predicted measurement residual variance	n	n	1
$K = vs$	K Kalman gain		n	
$\epsilon = z - a^{\mathrm{T}}\hat{x}$	ϵ Predicted residual $\hat{x}$ *A priori* estimate	n	n	
$\hat{x} := \hat{x} + K\epsilon$	The state update is the new *a priori* estimate	n	n	
$P := P - Kv^{\mathrm{T}}$	Covariance update (symmetric matrix)	$0.5n^2 + 0.5n$	$0.5n^2 + 0.5n$	
	Total number of operations to process m observations	$(1.5n^2 + 3.5n)m$	$(1.5n^2 + 4.5n)m$	m

[a] This mechanization corresponds to Eqs. (II.6.4), (II.6.12), and (II.6.13), and is given for reference purposes, but is not recommended for use when there is any danger of numerical difficulty.

Table B.2

Kalman Stabilized Data Processing[a]

$P := (I - Ka^{\mathrm{T}})P(I - Ka^{\mathrm{T}})^{\mathrm{T}} + KrK^{\mathrm{T}}$

Computation	Remarks		Number of additions	Number of multiplications	Number of divisions
$z := zr_{\nu}$	$r_{\nu} = 1/\sqrt{r}$			1	
$a^{\mathrm{T}} := a^{\mathrm{T}} r_{\nu}$	Scaling measurement			n	
$v = Pa$	v	Temporary work vector	n^2	n^2	
$s = 1/(a^{\mathrm{T}}v + 1)$	s^{-1}	Predicted measurement residual variance	n	n	1
$K = vs$	K	Kalman gain		n	
$\epsilon = z - a^{\mathrm{T}}\hat{x}$	ϵ $\hat{x}$	Predicted residual *A priori* estimate	n	n	
$\hat{x} := \hat{x} + K\epsilon$		The state update is the new *a priori* estimate	n	n	
$P := P - Kv^{\mathrm{T}}$	P	Used for temporary storage and computed as symmetric	$0.5n^2 + 0.5n$	$0.5n^2 + 0.5n$	
$v := Pa$			n^2	n^2	
$P := (P - vK^{\mathrm{T}}) + KK^{\mathrm{T}}$	Updated, symmetric covariance, off-diagonal elements averaged		$2n^2 + 2n$	$2n^2 + 2n$	
	Total number of operations to process m observations		$(4.5n^2 + 5.5n)m$	$(4.5n^2 + 7.5n)m$[b]	m

[a]The FORTRAN mechanization is given in Appendix II.D. This algorithm (cf. Appendix II.D) has better numeric characteristics than the standard formula of Table B.1.

[b]Multiplies and adds are included only up to order n.

Table B.3

Potter Square Root Covariance Update[a]

$S := S - \gamma K v^{\mathrm{T}}$

Computation	Remarks	Number of additions	Number of multiplications	Number of divisions	Number of square roots
$v = S^{\mathrm{T}} a$		n^2	n^2		
$\lambda = 1/(v^{\mathrm{T}} v + r)$	s^{-1} Predicted measurement residual variance	n	n	1	
$\epsilon = (z - a^{\mathrm{T}} \hat{x}) \lambda$	$\hat{x}$ *A priori* estimate; ϵ is $N(0, \lambda)$	n	n		
$\bar{K} = \mathbf{S} v$	Scaled Kalman gain	n^2	n^2		
$\hat{x} := \hat{x} + \bar{K}\epsilon$	State update	n	n		
$S := S - \left(\frac{\lambda}{1 + \sqrt{\lambda}} \bar{K} \right) v^{\mathrm{T}}$	Square root covariance update	n^2	$n^2 + n$	1	1
	Total[b] number of operations to process m observations	$(3n^2 + 3n)m$	$(3n^2 + 4n)m$	$2m$	m

[a] Potter's algorithm and its FORTRAN mechanization are given in Appendix II.E.

[b] Multiplies and adds are included only up to order n.

Table B.4
U–D Factorization Covariance Update[a]

Computation	Remarks	Number of additions	Number of multiplications	Number of divisions
$f = U^{\mathrm{T}}a$	U Unit upper triangular	$0.5n^2 - 0.5n$	$0.5n^2 - 0.5n$	
$v = Df$	D Diagonal $(d_1, \ldots, d_n)$		n	
$\alpha_j = \alpha_{j-1} + v_j f_j$	$\alpha_0 = r, j = 1, \ldots, n$, here and in the two equations to follow	n	n	
$\gamma_j = 1/\alpha_j$				n
$d_j := d_j \alpha_{j-1} \gamma_j$	Updated d factors		$2n$	
$\lambda_j = -f_j \gamma_j$	$j = 2, \ldots, n$ here and in the following three equations		n	
$\tilde{u}_j = u_j$	$K_1^{\mathrm{T}} = (v_1, 0, \ldots, 0)$			
$u_j := \tilde{u}_j + \lambda_j K_j$	$u_j :=$ jth column of U; only the first $j - 1$ elements need be set	$0.5n^2 - 0.5n$	$0.5n^2 - 0.5n$	
$K_{j+1} = K_j + v_j \tilde{u}_j$	Only the first j components of K_{j+1} are nonzero	$0.5n^2 - 0.5n$	$0.5n^2 - 0.5n$	
$\epsilon = (z - a^{\mathrm{T}}\hat{x})\gamma_n$	Computed residual, scaled	n	n	
$\hat{x} := \hat{x} + K_{n+1}\epsilon$	Updated estimate	n	n	
	Total[b] number of operations to process m observations	$(1.5n^2 + 1.5n)m$	$(1.5n^2 + 5.5n)m$	nm

[a]The algorithm is given in Theorem V.3.1 and the FORTRAN mechanization is given in Appendix V.A.
[b]Multiplies and adds are included only up to order n.

Table B.5
Upper Triangular Covariance Square Root Update[a]

Computation	Remarks	Number of additions	Number of multiplications	Number of divisions	Number of square roots
$\bar{f} = S^{\mathrm{T}}a$	S Upper triangular	$0.5n^2+0.5n$	$0.5n^2+0.5n$		
$\alpha_j = \alpha_{j-1}+\bar{f}_j^2$	$\alpha_0 = r;\ j = 1, \ldots, n$ here and in the equations to follow	n	n		
$\beta_j = (\alpha_{j-1}/\alpha_j)^{1/2}$	$0 < \beta_j \leqslant 1$			n	n
$S_{jj} := S_{jj}\beta_j$	Updated diagonals		n		
$\gamma_j = \bar{f}_j/(\beta_j\alpha_j)$	$j > 1$		n	$n-1$	
$\tilde{S}_{ij} = S_{ij}$					
$S_{ij} := S_{ij}\beta_j - \gamma_j K_i$	Updated off-diagonal terms $i = 1, \ldots, j-1$, and $j > 1$	$0.5n^2-0.5n$	n^2-n		
$K_i := K_i + \bar{f}_i\tilde{S}_{ij}$	$i = 1, \ldots, j$	$0.5n^2+0.5n$	$0.5n^2+0.5n$		
$\epsilon = (z - a^{\mathrm{T}}\hat{x})/\alpha_n$	Computed residual, scaled	n	n	1	
$\hat{x} := \hat{x} + K\epsilon$	Updated state estimate	n	n		
	Total[b] number of operations to process m observations	$(1.5n^2+3.5n)m$	$(2n^2+5n)m$	$2nm$	nm

[a]The algorithm is given as Corollary V.3.1 and the FORTRAN mechanization is given in Appendix V.A.

[b]Multiplies and adds are included only up to order n.

Table B.6
Normal Equation Data Accumulation[a]

Computation	Remarks	Number of additions	Number of multiplications	Number of divisions	Number of square roots
$z := zr_\nu$	Normalization of observation		1		
$a := ar_\nu$	$r_\nu = 1/\sqrt{r}$		n		
$\Lambda := \Lambda + a^{\mathrm{T}}a$	Λ The information matrix is symmetric	$0.5n^2+0.5n$	$0.5n^2+0.5n$		
$d := d + a^{\mathrm{T}}z$	$\hat{x} = \Lambda^{-1}d$	n	n		
	Total number of operations to accumulate m observations	$(0.5n^2 + 1.5n)m$	$(0.5n^2 + 2.5n)m$		
	For estimates and/or variances				
$RR^{\mathrm{T}} = \Lambda$	Compute Cholesky factors (cf. Appendix III.A)	$\frac{1}{6}n^3 - \frac{1}{6}n$	$\frac{1}{6}n^3 + 0.5n^2 - \frac{2}{3}n$	n	n
$R^{\mathrm{T}}y = d$ $R\hat{x} = y$	If estimates only wanted (cf. Eq. (IV.B.2))	n^2+n	n^2+n	n	
	Number of operations for estimates only	$\frac{1}{6}n^3 + n^2 + \frac{5}{6}n$	$\frac{1}{6}n^3 + 1.5n^2 + \frac{1}{3}n$	$2n$	n
$S = R^{-1}$	cf. Appendix IV.B	$\frac{1}{6}n^3 - \frac{1}{2}n^2 + \frac{1}{3}n$	$\frac{1}{6}n^3 + \frac{1}{2}n^2 - \frac{2}{3}n$	n	
$\hat{x} = R^{-\mathrm{T}}(R^{-1}d)$	This option used if estimates and covariance information are needed	n^2+n	n^2+n		
	Number of operations for estimates and square root covariance	$\frac{1}{3}n^3 + \frac{1}{2}n^2 + \frac{7}{6}n$	$\frac{1}{3}n^3 + 2n^2 - \frac{1}{3}n$	$2n$	n

[a]See Appendix II.C for a description of the alogrithm. One could also use the U–D factorization of Λ (cf. Section III.4) to compute estimates and variances.

Table B.7

SRIF Data Processing Using the Householder Algorithm with R Assumed Triangular[a]

Computation	Remarks	Number of additions	Number of multiplications	Number of divisions	Number of square roots
$A := A\ \mathrm{Diag}(r_v(1), \ldots, r_v(m))$	Scale measurement matrix to form normalized measurements		$m(n+1)$		
$\sigma := \mathrm{sgn}(R_{j,j})\left[R_{j,j}^2 + \sum_{k=1}^{m} a_{k,j}^2\right]^{1/2}$	Equation cycled through n times, $j = 1, \ldots, n$	$n(m+1)$	$n(m+1)$		n
$u := R_{j,j} + \sigma$	At the jth step the Householder u vector has in it u and the jth column of the current A matrix	n			
$R_{j,j} := \sigma$					
$\beta := 1/(\sigma u)$			n	n	

	$l = j + 1, \ldots, n + 1$ in the next three equations				
$\gamma := \beta \left(\sum_{k=1}^{m} a_{k,j} a_{k,l} + u R_{j,l} \right)$	The $(n+1)$st column of R is $\hat{z}$, and the $(n+1)$st column of A is z, i.e.,				
$R_{j,l} := R_{j,l} - \gamma u$					
$a_{i,l} := a_{i,l} - \gamma a_{i,j}, i = 1, \ldots, m$	$\begin{bmatrix} \hat{R} & \hat{z} \\ A & z \end{bmatrix} \to \begin{bmatrix} \hat{R} & \hat{z} \\ 0 & e \end{bmatrix}$	$(m+1)(n^2+n)$	$(2m+3)\, n(n+1)/2$		
	Total[b] number of operations for m observations	$n^2+3n+(n^2+2n)m$	$1.5n^2+3.5n$ $+(n^2+3n)m$	n	n

[a]The FORTRAN mechanization of this algorithm is given in Appendix VII.B, and the algorithm is described in Appendix IV.A.
[b]Multiplies and adds are included only up to order n.

References

[1] Carlson, N. A., Fast triangular factorization of the square root filter, *AIAA J.* **11**, No. 9, 1259–1265 (1973).

The key ideas of this paper are the triangular factorization of the optimal Kalman updating algorithm and the arrangement of results so that numerical differencing is avoided. We capitalized on these ideas in Section V.3 when developing the *U–D* algorithm.

[2] Lawson, C. L., and Hanson, R. J., "Solving Least Squares Problems." Prentice-Hall, Englewood Cliffs, New Jersey (1974).

This book elaborates and expands on the material in Section V.2 on the SRIF data processing algorithm. The authors' extensive numerical experience is evidenced by the extensive collection of examples that are included and by the accuracy enhancement techniques that they recommend to offset the various ill-conditionings that arise in practice.

[3] Kaminski, P. G., Bryson, A. E., and Schmidt, S. F., Discrete square root filtering: A survey of current techniques, *IEEE Trans. Automatic Contr.* **AC-16**, No. 6, 727–735 (1971).

This survey, based to a large extent on Kaminski's thesis, summarizes the then state of the art in regard to filtering. Recent advances in this area are discussed in this book.

[4] Bierman, G. J., A comparison of discrete linear filtering algorithms, *IEEE Trans. Aero. Elect. Systems* **AES-9**, No. 1, 28–37 (1973).

This filter algorithm survey is, in the main, an elaboration on the survey given by Kaminski *et al.* [3]. The material of this survey supplements the material presented in Section V.4.

[5] Kaminski, P. G., Square root filtering and smoothing for discrete processes, Ph.D. Dissertation, Dept. Aeronaut. Astronaut., Stanford Univ., Stanford, California (September 1971).

The main contributions of this well written thesis are the examples which highlight the value of using square root algorithms and the backward square root smoothing algorithm which elegantly solves the discrete smoothing problem. This material supplements Sections V.4 and V.5.

[6] Thornton, C. L., and Bierman, G. J., A numerical comparison of discrete Kalman filtering algorithms: an orbit determination case study. JPL Tech. Memo. 33–771 (June 1976).

This thorough orbit determination case study illustrates the impact that numerical errors have on estimation accuracy. The results demonstrate a variety of numerical failings of Kalman-type covariance algorithms. Factorization algorithms are shown to be relatively impervious to numerical errors.

VI

Inclusion of Mapping Effects and Process Noise

VI.1 Introduction

In Chapters II and V attention was restricted to parameter estimation, i.e., we restricted our discussions to least squares estimation of parameter vectors that do not vary with time. Suppose now that

$$x_{j+1} = \Phi_j x_j + G w_j \tag{1.1}$$

where $\Phi_j = \Phi(t_{j+1}, t_j)$ is the nonsingular transition matrix relating the state at time t_j to the state at time t_{j+1}, and w_j is a p vector of *process noise* parameters that are assumed to have a nonsingular covariance matrix[†] Q. In general, p, G, and Q are j dependent, but this dependence is suppressed for notational convenience.

Remark: The vector x_j represents the state of the system at time t_j and should not be confused with the previous notation $\hat{x}_j$ (which represented estimates of the constant vector x with data from j batches).[‡] The "^" notation, to be discussed again later in this section, is consistent with our previous notation[§] if we interpret the previous problem with j batches to contain j steps of Eq. (1.1) with $\Phi = I$ and $G = 0$.

The process noise w_j sometimes represents the effects of unmodeled parameters, or errors due to linearization, and sometimes it accounts for

[†]Note that there is no loss of generality in assuming that Q is nonsingular. Any problem with a singular w covariance could be converted to a nonsingular problem by suitably redefining G.

[‡]One often speaks of vector observations as batches of measurements or simply as *batches*.

[§]Occasional notational inconsistencies appear, where subscripting represents components of a vector or matrix. There should, however, be no confusion in these cases because the context should prevent any misunderstanding.

effects such as leaky attitude controls or solar winds. The mathematical description of process noise is that it is a random process with

$$E(w_j) = \bar{w}_j \qquad \text{and} \qquad E\left[(w_j - \bar{w}_j)(w_k - \bar{w}_k)^{\mathrm{T}}\right] = Q\,\delta_{j,k} \tag{1.2}$$

where $\delta_{j,k}$ is the Kronecker delta function, which is zero except when $j = k$.

In almost all of our applications it happens that the *a priori* estimate $\bar{w}_j$ is zero; however, we do not assume this in our development. By assumption, $Q > 0$, so that it can be written as

$$Q = R_w^{-1} R_w^{-\mathrm{T}} \tag{1.3}$$

In this chapter we set about estimating the vectors x_j from a set of linear observations:

$$z_j = A_j x_j + \nu_j, \qquad j = 0, \ldots, N \tag{1.4}$$

where $\nu_j \in N(0, I)$.†

In many texts and papers one encounters the notation $x_{k/N}$ as the estimate of x_k based on data up to and including time N. Corresponding to the estimate there is a covariance matrix, denoted $P_{k/N}$,

$$P_{k/N} = E\left[(x_k - x_{k/N})(x_k - x_{k/N})^{\mathrm{T}}\right]$$

When $k > N$ the estimate $x_{k/N}$ is said to be the *predicted* estimate, i.e., we are trying to predict where the state will be at some future time t_k based on data up to some earlier time t_N. When $k = N$, $x_{k/k}$ is a *filtered* estimate and we are estimating the current state based on data up to the present time. Finally, when $k < N$ we say that $x_{k/N}$ is the *smoothed* estimate of x_k, i.e., we want to reconstruct past states.

The double subscript notation becomes unwieldly when the component elements of x and/or P are involved. Indeed, in these cases one becomes enmeshed in expressions that involve three or four indices. The only subscript pairs we shall have occasion to deal with are $x_{j/j-1}$, the single-stage predicted estimate of x_j using data up to and including time t_{j-1}; $x_{j/j}$, the filter estimate of x_j using data up to and including time t_j; and $x_{j/N}$, the smoothed estimate of x_j based on all the data up to and including the terminal time t_N, $N > j$. Notation used for these cases is ~, ^, and $*$, viz.,

$$x_{j/j-1} \equiv \tilde{x}_j, \qquad x_{j/j} \equiv \hat{x}_j, \qquad \text{and} \qquad x_{j/N} \equiv x_j^* \tag{1.5}$$

† $\nu \in N(0, I)$ is an abbreviation for $E(\nu) = 0$ and $\operatorname{cov}(\nu) = I$. It also means that ν is a random vector with a normal distribution, but we do not make use of this.

When dealing with parameter estimation, $x_j = x_{j-1}$, and therefore the predicted estimate at time t_j is the same as the filtered estimate at time t_{j-1} (i.e., $\tilde{x}_j = \hat{x}_{j-1}$). This explains the initialization of recursion (V.2.8); the predicted (*a priori*) information at the start of the problem coincides with the filtered results before any data have been included.

Remark: The notation, ~, ^, and * will also be applied to variables such as R and z. For example,

$$\hat{P} = \hat{R}^{-1}\hat{R}^{-\mathrm{T}} \qquad \text{and} \qquad \hat{z} = \hat{R}\hat{x}$$

In this chapter attention is focused on the filtering problem. We first developed a square root filtering algorithm that arises when minimizing a suitable sum of squares performance functional. This development (Section VI.2) is a simple extension of the square root data processing method discussed in Section V.2. In essence, the extension consists of modifying the sum-of-squares performance functional to include process noise, and then partitioning the result.

In Chapter II we developed the Kalman data processing algorithm as an alternative data processing method. This algorithm is simply implemented and is well suited to problems that involve frequent estimates and/or covariances. In Section VI.3 we will develop equations that recursively propagate the state estimate and its covariance (from time t_j to time t_{j+1}).

The U–D covariance factorization that was introduced in the last chapter is extended in Section VI.4 to include mapping and process noise effects. The chapter concludes with a modest discussion of covariance-information duality relations. This notion is used to derive a triangular square root covariance algorithm that is compatible with Potter's covariance result and with the triangular factorization update algorithm of Appendix V.A.

VI.2 Mapping and the Inclusion of Process Noise into the SRIF

The principle of the SRIF mapping and process noise inclusion method can be explained by considering a problem with a single time step. Consider the problem of estimating x_1 given

$$z_0 = A_0 x_0 + \nu_0 \tag{2.1}$$

and

$$x_1 = \Phi_0 x_0 + G w_0 \tag{2.2}$$

with *a priori* information given about x_0 and w_0. We suppose that the *a priori* information on x_0 and w_0 is given in data equation form:

$$z_w = R_w w_0 + \nu_w \tag{2.3}$$

$$\tilde{z}_0 = \tilde{R}_0 x_0 + \tilde{\nu}_0 \tag{2.4}$$

The ν variables ν_0, $\tilde{\nu}_0$, and ν_w are assumed to be zero mean, independent, and to have unity covariances. We choose as our performance functional $J^{(1)}$, the *likelihood functional*,

$$J^{(1)} = \|\tilde{z}_0 - \tilde{R}_0 x_0\|^2 + \|z_0 - A_0 x_0\|^2 + \|z_w - R_w w_0\|^2 \tag{2.5}$$

The x_0 and w_0 that minimize the likelihood functional (2.5), together with the mapping equation (2.2), give the x_1 estimate.

Observe that (2.5) is a sum of nonnegative terms and that the first two terms correspond to the least squares problem of estimating x_0 so as to minimize the residual measurement errors. Thus drawing upon the material of Chapter V, we write

$$J^{(1)} = \left\| \begin{bmatrix} \tilde{R}_0 \\ A_0 \end{bmatrix} x_0 - \begin{bmatrix} \tilde{z}_0 \\ z_0 \end{bmatrix} \right\|^2 + \|R_w w_0 - z_w\|^2$$

$$= \left\| \begin{bmatrix} \hat{R}_0 \\ 0 \end{bmatrix} x_0 - \begin{bmatrix} \hat{z}_0 \\ e_0 \end{bmatrix} \right\|^2 + \|R_w w_0 - z_w\|^2 \qquad \text{(cf. Eq. (V.2.8))}$$

$$= \|e_0\|^2 + \|\hat{R}_0 x_0 - \hat{z}_0\|^2 + \|R_w w_0 - z_w\|^2 \tag{2.6}$$

To express $J^{(1)}$ in terms of x_1 we eliminate x_0 from (2.6):

$$J^{(1)} = \|e_0\|^2 + \|\hat{R}_0 \Phi_0^{-1}(x_1 - G w_0) - \hat{z}_0\|^2 + \|R_w w_0 - z_w\|^2$$

$$= \|e_0\|^2 + \left\| \begin{bmatrix} R_w & 0 \\ -R_1^{\mathrm{d}} G & R_1^{\mathrm{d}} \end{bmatrix} \begin{bmatrix} w_0 \\ x_1 \end{bmatrix} - \begin{bmatrix} z_w \\ \hat{z}_0 \end{bmatrix} \right\|^2 \tag{2.7}$$

where $R_1^{\mathrm{d}} = \hat{R}_0 \Phi_0^{-1}$.

Equation (2.7) is an improvement over the original problem formulation because now we no longer need to compute x_0 prior to computing x_1. This point is important, because in the general problem it would be undesirable to have to compute (smoothed) estimates of all the previous states prior to obtaining the filtered estimate of the current state. Note that $\hat{x}_1$, the minimizing value of x_1 in (2.7), still depends on w_0. Projecting this observation to the general case, one can conjecture that it would be undesirable to have to solve for $w_0, \ldots, w_{n-1}$ before being able to obtain x_n. A method of avoiding this explicit dependence of x_1 on w_0 is to replace the matrix equation of (2.7) by an (orthogonally) equivalent equation which eliminates w_0 from all but the first N_w equations (N_w being the dimension of w). Thus an orthogonal transformation $\tilde{T}_1$ is selected (cf. Chapter IV) such that

$$\tilde{T}_1 \left[\begin{array}{c|c|c} \overbrace{R_w}^{N_w} & \overbrace{0}^{N_x} & \overbrace{z_w}^{1} \\ \hline -R_1^d G & R_1^d & \hat{z}_0 \end{array}\right] \begin{matrix} \}N_w \\ \}N_x \end{matrix} = \left[\begin{array}{c|c|c} \overbrace{\tilde{R}_w(1)}^{N_w} & \overbrace{\tilde{R}_{wx}(1)}^{N_x} & \overbrace{\tilde{z}_w(1)}^{1} \\ \hline 0 & \tilde{R}_1 & \tilde{z}_1 \end{array}\right] \begin{matrix} \}N_w \\ \}N_x \end{matrix} \tag{2.8}$$

where $\tilde{T}_1$, a product of N_w elementary Householder transformations, partially triangularizes the matrix. Replacing the normed matrix equation in (2.7) by the terms of (2.8) gives the result

$$\begin{aligned} J^{(1)} &= \|e_0\|^2 + \left\| \begin{bmatrix} \tilde{R}_w(1) & \tilde{R}_{wx}(1) \\ 0 & \tilde{R}_1 \end{bmatrix} \begin{bmatrix} w_0 \\ x_1 \end{bmatrix} - \begin{bmatrix} \tilde{z}_w(1) \\ \tilde{z}_1 \end{bmatrix} \right\|^2 \\ &= \|e_0\|^2 + \|\tilde{R}_w(1)w_0 + \tilde{R}_{wx}(1)x_1 - \tilde{z}_w(1)\|^2 + \|\tilde{R}_1 x_1 - \tilde{z}_1\|^2 \end{aligned} \tag{2.9}$$

Because orthogonal transformations are rank preserving it follows that $\tilde{R}_w(1)$ is nonsingular.† Thus it is always possible to choose $w_0 = \tilde{w}_0$ so that the middle term of (2.9) is zero, where

$$\tilde{w}_0 = \left[\tilde{R}_w(1)\right]^{-1}\left(\tilde{z}_w(1) - \tilde{R}_{wx}(1)\tilde{x}_1\right) \tag{2.10}$$

†To see this clearly, note that

$$\tilde{T}_1 \left[\begin{array}{c} \overbrace{R_w}^{N_w} \\ \hline -R_1^d G \end{array}\right] \}N_w = \left[\begin{array}{c} \tilde{R}_w(1) \\ \hline 0 \end{array}\right]$$

The right side has full rank since R_w, on the left, is nonsingular.

Since R_1^{d} is nonsingular, this property is transmitted to $\tilde{R}_1$ via the transformation $\tilde{T}_1$. Thus the last term of (2.9) can also be made to vanish, and the minimum value of $J^{(1)}$ is $\|e_0\|^2$.

From (2.9) we infer that the information array for x_1 is $[\tilde{R}_1\ \tilde{z}_1]$, i.e.,

$$\tilde{x}_1 = \tilde{R}_1^{-1}\tilde{z}_1 \qquad \text{and} \qquad \tilde{P}_1 = E\left[(x_1 - \tilde{x}_1)(x_1 - \tilde{x}_1)^{\mathrm{T}}\right] = \tilde{R}_1^{-1}\tilde{R}_1^{-\mathrm{T}}$$

The result just described is in essence the SRIF solution to the general problem of combining *a priori* information, data, and mapping. To further clarify the basic ideas involved and to highlight the key mathematical steps, we shall derive the result again; this time using arguments that are analogous to those used in the previous chapter on square root information data processing.

From (2.5) and (2.6) it can be seen that the first step of the filtering algorithm consists of combining the *a priori* in data equation form, Eq. (2.3), with the data at time t_0, Eq. (2.1), to form an updated data equation (cf. Eq. (2.6) and Chapter V)). Since we already know how to combine data and *a priori* using data arrays, we focus our attention on the problem of combining the mapping (2.2) with the updated *a priori* data equation for x_0:

$$\hat{R}_0 x_0 = \hat{z}_0 - \hat{\nu}_0 \tag{2.11}$$

and the *a priori* data equation for w_0, Eq. (2.3). Our goal is the formation of an information array $[\tilde{R}_1\ \tilde{z}_1]$ for x_1.

If only the estimate and covariance of x_1 were desired, one could obtain them in the following manner:

$$\hat{x}_0 = \hat{R}_0^{-1}\hat{z}_0, \qquad \tilde{w}_0 = R_w^{-1}z_w \qquad \text{(cf. Eq. (2.6))} \tag{2.12}$$

$$\tilde{x}_1 = \Phi_0\hat{x}_0 + G\tilde{w}_0 \tag{2.13}$$

To obtain the covariance of $\tilde{x}_1$, first form the error equation

$$x_1 - \tilde{x}_1 = \Phi_0(x_0 - \hat{x}_0) + G(w_0 - \tilde{w}_0) \tag{2.14}$$

by differencing (2.13) and (2.2).

Combining (2.3), (2.11), and (2.12), we obtain

$$x_0 - \hat{x}_0 = -\hat{R}_0^{-1}\hat{\nu}_0, \qquad w_0 - \tilde{w}_0 = -R_w^{-1}\nu_w$$

Thus (2.14) can be rewritten as

$$x_1 - \tilde{x}_1 = -\Phi_0\hat{R}_0^{-1}\hat{\nu}_0 - GR_w^{-1}\nu_w$$

Since $\hat{\nu}_0$ and ν_w are zero mean, independent, and have unit covariance, we have

$$\tilde{P}_1 = E\left[(x_1 - \tilde{x}_1)(x_1 - \tilde{x}_1)^{\mathrm{T}}\right] = \left(\Phi_0 \hat{R}_0^{-1}\right)\left(\Phi_0 \hat{R}_0^{-1}\right)^{\mathrm{T}} + GR_w^{-1}\left(GR_w^{-1}\right)^{\mathrm{T}}$$

$$= \Phi_0 \hat{P}_0 \Phi_0^{\mathrm{T}} + GQG^{\mathrm{T}} \tag{2.15}$$

We shall return to the estimate covariance pair (2.13) and (2.15) when we discuss the Kalman filter in Section VI.3.

Let us define

$$x_1^{\mathrm{d}} = \Phi_0 x_0 \tag{2.16}$$

so that x_1^{d} is the deterministic mapping,† i.e., it is that part of the mapping that is independent of the process noise. Substituting (2.16) into the *a priori* equation (2.11), we obtain

$$R_1^{\mathrm{d}} x_1^{\mathrm{d}} = \hat{z}_0 - \hat{\nu}_0 \tag{2.17}$$

where $R_1^{\mathrm{d}} = \hat{R}_0 \Phi_0^{-1}$.

One can think of a predicted estimate at t_1:

$$\hat{x}_1^{\mathrm{d}} = \Phi_0 \hat{x}_0 \tag{2.18}$$

and

$$\hat{P}_1^{\mathrm{d}} = \Phi_0 \hat{P}_0 \Phi_0^{\mathrm{T}} \qquad \text{(predicted mapped covariance)}$$

$$= \left(R_1^{\mathrm{d}}\right)^{-1}\left(R_1^{\mathrm{d}}\right)^{-\mathrm{T}} \tag{2.19}$$

Equations (2.18) and (2.19) would be the least squares estimate and covariance at time t_1 if there were no process noise. Model (2.2), however, shows that

$$x_1 = x_1^{\mathrm{d}} + Gw_0 \tag{2.20}$$

and thus, to obtain a data equation for x_1, one should substitute (2.20) into (2.17):

$$R_1^{\mathrm{d}}(x_1 - Gw_0) = \hat{z}_0 - \hat{\nu}_0$$

i.e.,

$$-R_1^{\mathrm{d}} G w_0 + R_1^{\mathrm{d}} x_1 = \hat{z}_0 - \hat{\nu}_0 \tag{2.21}$$

†Strictly speaking, x_1^{d} is not deterministic since x_0 is random.

This equation is not in the form of a data equation; when, however, the *a priori* w_0 data equation (2.3) is addended we obtain

$$\begin{bmatrix} R_w & 0 \\ -R_1^{\mathrm{d}}G & R_1^{\mathrm{d}} \end{bmatrix} \begin{bmatrix} w_0 \\ x_1 \end{bmatrix} = \begin{bmatrix} z_w \\ \hat{z}_0 \end{bmatrix} - \begin{bmatrix} \nu_w \\ \hat{\nu}_0 \end{bmatrix} \tag{2.22}$$

This is the equation encountered in (2.7). It is a data equation for the augmented vector

$$\begin{bmatrix} w_0 \\ x_1 \end{bmatrix}$$

It is not exactly what we were looking for; we wanted a data equation for x_1. To manufacture the desired data equation we recall that in Section V.2 it was remarked that data equations are nonunique and that an orthogonal transformation maps a data equation into an equivalent data equation (i.e., both give the same estimates and covariances). We exploit this property by choosing the transformation $\tilde{T}_1$ to partially triangularize the coefficient matrix of (2.22). This partial triangularization eliminates w_0 from the last N_x rows of (2.22), and the result is Eq. (2.8). It is the mechanism by which mapping and process noise are included into the SRIF.

To see the significance of the right side of (2.8), we write the associated data equations

$$\tilde{R}_w(1)w_0 + \tilde{R}_{wx}(1)x_1 = \tilde{z}_w(1) - \tilde{\nu}_w \tag{2.23}$$

$$\tilde{R}_1 x_1 = \tilde{z}_1 - \tilde{\nu}_1 \tag{2.24}$$

The first equation (2.23) is a relation between the process noise w_0 and the state x_1. This equation will be the cornerstone of our later smoothing applications because it enables one to calculate a smoothed estimate of the process noise at time t_0 from a smoothed estimate of the state at time t_1. The second equation (2.24) is the sought after data equation; from it one can construct an estimate and/or covariance, if desired. More importantly, the form of (2.24) allows it to be used as an *a priori* for the next mapping step.

The augmented state vector appearing in (2.22) specifies w_0 first because the orthogonal transformation to be applied (cf. the matrix triangularization theorem of Chapter IV) is an elimination method. The first component $w_0(1)$ is eliminated from all equations below the first; the second component $w_0(2)$ is eliminated from all equations below the second, etc. This technique is analogous to the Gaussian elimination method of solving a system of linear equations. The reason we use orthogonal transformations instead of simple Gausian elimination is that the latter method does not preserve the unit covariance property of the noise terms in the data equation.

By this time the reader may be speculating that since the problem of computing the information array at time t_1 required such a lengthy explanation, the general problem, at time t_N, must be insurmountable. Fortunately, this is not the case and the subscripts zero and one represent the general case from time t_j to time t_{j+1}. To see that this is indeed the case, consider the general performance functional $J^{(N+1)}$

$$J^{(N+1)} = \|\tilde{R}_0 x_0 - \tilde{z}_0\|^2 + \sum_{j=0}^{N} \left(\|A_j x_j - z_j\|^2 + \|R_w(j) w_j - z_w(j)\|^2\right) \tag{2.25}$$

$$= J^{(N)} + \|A_N x_N - z_N\|^2 + \|R_w(N) w_N - z_w(N)\|^2 \tag{2.26}$$

From (2.26) and (2.9) it is easy to show that

$$J^{(N+1)} = \sum_{j=0}^{N} \left(\|e_j\|^2 + \|\tilde{R}_w(j+1) w_j + \tilde{R}_{wx}(j+1) x_{j+1} - \tilde{z}_w(j+1)\|^2\right) + \|\tilde{R}_{N+1} x_{N+1} - \tilde{z}_{N+1}\|^2 \tag{2.27}$$

where the e and $\tilde{R}$ terms are generated via the SRIF recursions (2.28) and (2.29).

Data Processing

$$\hat{T}_j \begin{bmatrix} \tilde{R}_j & \tilde{z}_j \\ A_j & z_j \end{bmatrix} = \begin{bmatrix} \hat{R}_j & \hat{z}_j \\ 0 & e_j \end{bmatrix} \tag{2.28}$$

Mapping

$$\tilde{T}_{j+1} \left[\begin{array}{c|c|c} R_w(j) & 0 & z_w(j) \\ -R^{\mathrm{d}}_{j+1} G & R^{\mathrm{d}}_{j+1} & \hat{z}_j \end{array}\right] = \left[\begin{array}{c|c|c} \tilde{R}_w(j+1) & \tilde{R}_{wx}(j+1) & \tilde{z}_w(j+1) \\ 0 & \tilde{R}_{j+1} & \tilde{z}_{j+1} \end{array}\right] \tag{2.29}$$

$$R^{\mathrm{d}}_{j+1} = \hat{R}_j \Phi^{-1}_{j+1}$$

Remark: The right-hand side of the first row of (2.29) is not used in filtering. It will, however, be used to solve the smoothing problem.

In practice the R_w matrices are diagonal. Indeed, if the original problem involved a nondiagonal noise covariance Q, one could transform the problem to a diagonal form by noting that $Gw_j = GR_w^{-1}(R_w w_j)$ and that the vector $R_w w_j$ has unity covariance.† In Chapter VII we delve into the computational aspects of the filtering problem, and there we will explain the advantages of evaluating the mapping equation (2.29) by including the process noise one component at a time.

VI.3 Mapping and the Inclusion of Process Noise into the Kalman Filter

In Chapter II we developed the Kalman data processing algorithm by seeking a linear, unbiased, minimum variance estimator. It turned out that this estimator was the same as the least squares estimator. In the case of mapping and the inclusion of process noise it is also true that the Kalman estimator is the same as the least squares estimator. The requirement that the estimate be unbiased implies (cf. Eq. (1.1))

$$\tilde{x}_{j+1} = \Phi_j \hat{x}_j + G\tilde{w}_j \tag{3.1}$$

and the covariance of this estimate is determined from the error equation (cf. Eq. (2.14))

$$x_{j+1} - \tilde{x}_{j+1} = \Phi_j(x_j - \hat{x}_j) + G(w_j - \tilde{w}_j)$$

The covariance follows from this equation

$$\tilde{P}_{j+1} = E\left[(x_{j+1} - \tilde{x}_{j+1})(x_{j+1} - \tilde{x}_{j+1})^{\mathrm{T}}\right] = \Phi_j \hat{P}_j \Phi_j^{\mathrm{T}} + GQG^{\mathrm{T}} \tag{3.2}$$

It is interesting to note that the mapping equations are determined by the requirement that the estimator be unbiased.

Equations (3.1) and (3.2) show how the estimate and covariance at time t_j are mapped to time t_{j+1}, where they will (quite likely) be used as the *a priori* statistics and combined with the measurements at time t_{j+1}. The

†Note that representing the process noise in this way involves replacing G by $GQ^{1/2}$ (and using normalized, unit covariance process noise). A benefit accruing from this representation is that one can use a single equation mechanization that applies to both the process noise problem and the simple deterministic mapping (corresponding to $Q/= 0$). Nishimura's square root implementation of the Kalman filter error analysis program uses this concept (cf. Nishimura and Nead (1973)).

Kalman filter consists of the data processing algorithm (cf. Appendix II.D), together with the estimate and covariance recursions, (3.1) and (3.2).

It is easy to show that the estimate covariance calculation, using (3.1) and (3.2), is the same as the SRIF mapping (2.29). To demonstrate this point we partition the orthogonal transformation $\tilde{T}_{j+1}$

$$\tilde{T}_{j+1} = \begin{array}{c} \overbrace{\quad}^{N_w} \; \overbrace{\quad}^{N_x} \\ \left[\begin{array}{c|c} \tilde{T}_{11} & \tilde{T}_{12} \\ \hline \tilde{T}_{21} & \tilde{T}_{22} \end{array}\right] \end{array} \begin{array}{l} \}N_w \\ \}N_x \end{array} \tag{3.3}$$

and expand (2.29). Examining the bottom row of that matrix equation gives:

$$-\tilde{T}_{22}R^{\mathrm{d}}_{j+1}G + \tilde{T}_{21}R_w = 0 \tag{3.4}$$

$$\tilde{T}_{22}R^{\mathrm{d}}_{j+1} = \tilde{R}_{j+1} \tag{3.5}$$

$$\tilde{T}_{22}\hat{z}_j + \tilde{T}_{21}z_w(j) = \tilde{z}_{j+1} \tag{3.6}$$

If we now make use of the orthogonality of $\tilde{T}_{j+1}$ ($\tilde{T}_{j+1}\tilde{T}^{\mathrm{T}}_{j+1} = I$), we obtain the additional relation

$$\tilde{T}_{21}\tilde{T}^{\mathrm{T}}_{21} + \tilde{T}_{22}\tilde{T}^{\mathrm{T}}_{22} = I \tag{3.7}$$

The covariance result now follows by solving (3.4) for $\tilde{T}_{21}$, substituting it into (3.7), and after rearranging terms, using (3.5). The equations make the process self-explanatory:

$$\tilde{T}_{21} = \tilde{T}_{22}R^{\mathrm{d}}_{j+1}GR_w^{-1}$$

$$\begin{aligned} I &= \tilde{T}_{22}\left[I + R^{\mathrm{d}}_{j+1}GR_w^{-1}(R^{\mathrm{d}}_{j+1}GR_w^{-1})^{\mathrm{T}}\right]\tilde{T}^{\mathrm{T}}_{22} \\ &= \tilde{T}_{22}R^{\mathrm{d}}_{j+1}\left[(R^{\mathrm{d}}_{j+1})^{-1}(R^{\mathrm{d}}_{j+1})^{-\mathrm{T}} + GR_w^{-1}R_w^{-\mathrm{T}}G^{\mathrm{T}}\right](\tilde{T}_{22}R^{\mathrm{d}}_{j+1})^{\mathrm{T}} \\ &= \tilde{R}_{j+1}[P^{\mathrm{d}}_{j+1} + GQG^{\mathrm{T}}]\tilde{R}^{\mathrm{T}}_{j+1} \end{aligned}$$

Therefore

$$\tilde{R}^{-1}_{j+1}\tilde{R}^{-\mathrm{T}}_{j+1} = P^{\mathrm{d}}_{j+1} + GQG^{\mathrm{T}}$$

and this result is the same as (3.2) where $P^{\mathrm{d}}_{j+1} = \Phi_j \hat{P}_j \Phi_j^{\mathrm{T}}$ and $\tilde{R}_{j+1}^{-1}\tilde{R}_{j+1}^{-\mathrm{T}} = \tilde{P}_{j+1}$. The equivalence of (3.6) with estimate (3.1) is obtained similarly and the equations demonstrating this are

$$\tilde{z}_{j+1} = \tilde{T}_{22}\hat{z}_j + \tilde{T}_{21}z_w(j) \qquad \text{(Eq. (3.6))}$$

$$= \tilde{T}_{22}\hat{R}_j\hat{x}_j + \tilde{T}_{21}R_w\overline{w}_j \qquad \text{(cf. Eq. (2.12))}$$

$$= \tilde{T}_{22}\hat{R}_j\hat{x}_j + \tilde{T}_{22}R^{\mathrm{d}}_{j+1}G\overline{w}_j \qquad \text{(cf. Eq. (3.4))}$$

$$= \tilde{T}_{22}R^{\mathrm{d}}_{j+1}\left[\left(R^{\mathrm{d}}_{j+1}\right)^{-1}\hat{R}_j\hat{x}_j + G\overline{w}_j\right]$$

$$= \tilde{R}_{j+1}[\Phi_j\hat{x}_j + G\overline{w}_j] \qquad \left(\text{cf. Eq. (3.5) and the definition of } R^{\mathrm{d}}_{j+1}\right)$$

$$= \tilde{R}_{j+1}\tilde{x}_{j+1} \qquad \text{(mapping equation (1.1))}$$

Thus we have shown that the SRIF mapping (2.29), and the Kalman filter mapping (3.1)–(3.2) are algebraically equivalent.†

VI.4 Mapping and the Inclusion of Process Noise into the U–D Covariance Filter

Attention in this section is focused on updating the *U–D* covariance factors (cf. Section V.3), which correspond to the covariance time update equation (3.2). For notational simplicity the *j* subscript is omitted, and we write simply

$$\tilde{P} = \Phi\hat{P}\Phi^{\mathrm{T}} + GQG^{\mathrm{T}} \tag{4.1}$$

Thus given that $\hat{P} = \hat{U}\hat{D}\hat{U}^{\mathrm{T}}$, we seek the factors $\tilde{U}$ and $\tilde{D}$ such that $\tilde{P} = \tilde{U}\tilde{D}\tilde{U}^{\mathrm{T}}$, where the U factors are unit upper triangular and the D factors are diagonal. Provided that there is no danger of numeric accuracy deterioration, one could, in a brute force fashion, compute $\tilde{P}$ and then apply the *U–D* factorization algorithm which is given in Appendix III.A.

†For certain special cases the SRIF and Kalman algorithms may not be defined. For example, the SRIF algorithm cannot accommodate problems with perfect *a priori* knowledge (zero variances), while the Kalman covariance algorithm cannot accommodate problems involving no *a priori* knowledge (infinite variances).

We prefer instead to use a modified Gram–Schmidt type procedure (cf. Björck (1967) and Lawson and Hanson (1974)), to compute the updated factors.† This method, the principal subject of this section, is competitive with the direct method in terms of computational efficiency. Furthermore, the modified Gram–Schmidt triangularization is reputed (cf. Björck (1967) and Jordan (1968)) to have accuracy that is comparable with the Householder method.

We assume that the problem has been formulated so that Q is diagonal‡ and we factor (4.1) as follows:

$$\tilde{P} = [\Phi\hat{U} \quad G] \operatorname{Diag}(\hat{D}, Q) [\Phi\hat{U} \quad G]^{\mathrm{T}}$$

i.e.,

$$\tilde{P} = WDW^{\mathrm{T}} \tag{4.2}$$

where

$$W = [\overbrace{\Phi\hat{U}}^{n} \quad \overbrace{G}^{N_w}]\}n = \begin{bmatrix} w_1^{\mathrm{T}} \\ \vdots \\ w_n^{\mathrm{T}} \end{bmatrix}, \qquad n = N_x \tag{4.3}$$

and

$$D = \operatorname{Diag}(\hat{D}, Q) = \operatorname{Diag}(D_1, \ldots, D_N), \qquad N = n + N_w \tag{4.4}$$

The Gram–Schmidt process we have in mind corresponds to orthogonalizing the vectors $w_1, \ldots, w_n$ with respect to the weighted norm $\|w_j\|^2 \triangleq \sum_{k=1}^{N} D_k w_j(k)^2$.§ The algorithm Eqs. (4.5) and (4.6), for orthogonalizing a set of vectors with respect to this weighted norm will be called the weighted Gram–Schmidt (WG–S) algorithm.

Lemma VI.4.1 (Weighted Gram–Schmidt Orthogonalization) Let $w_j, \ldots, w_n$ be linearly independent N vectors with $N \geqslant n$ and let $D_1, \ldots, D_N$ be positive scalars with $D = \operatorname{Diag}(D_1, \ldots, D_N)$. If v_n,

†The idea of using a weighted Gram–Schmidt orthogonalization to time update the U–D covariance factors is due to Thornton (1976).

‡As a practical matter Q is generally diagonal, but if not, then factor it (cf. Appendix III.A) as $Q = U_q D_q U_q^{\mathrm{T}}$ and replace G and Q by GU_q and D_q, respectively. The reason for such a substitution is that $GQG^t = (GU_q)D_q(GU_q)^{\mathrm{T}}$.

§Recall that in our notation $w_j^{\mathrm{T}} = (w_j(1), \ldots, w_j(N))$.

$v_{n-1}, \ldots, v_1$ are defined by (4.5) and (4.6), then none of the v's are zero and $v_k^T D v_j = 0$ for $k \neq j$:

$$v_n = w_n \tag{4.5}$$

$$v_j = w_j - \sum_{k=j+1}^{n} \left(w_j^T D v_k / v_k^T D v_k \right) v_k, \qquad j = n-1, \ldots, 1 \tag{4.6}$$

Proof: This result (although not stated exactly as we have it) is one of the cornerstones of linear algebra. The proof, like the algorithm, is given recursively.

The algorithm is initialized using (4.5) with a nonzero v_n, $v_n \neq 0$, because of the hypothesis that the w's are linearly independent. Now $v_n \neq 0$ implies that $v_n^T D v_n > 0$ (because $D > 0$). This is important because $v_{n-1}, \ldots, v_1$ all involve division by $v_n^T D v_n$. To prove the orthogonality that is claimed, premultiply Eq. (4.6) by $v_l^T D$

$$v_l^T D v_j = v_l^T D w_j - \sum_{k=j+1}^{n} w_j^T D v_k \left(v_l^T D v_k / v_k^T D v_k \right) \tag{4.7}$$

For $j = n - 1$ and $l = n$, Eq. (4.7) shows that $v_n^T D v_{n-1} = 0$. Further, v_{n-1} is not zero because its construction guarantees that it is a nontrivial[†] linear combination of the linearly independent vectors w_{n-1} and w_n. Note $v_{n-1} \neq 0$ implies that $v_{n-1}^T D v_{n-1} > 0$ and this is important because $v_{n-2}, \ldots, v_1$ all involve division by $v_{n-1}^T D v_{n-1}$.

For $j < n - 1$ we use Eq. (4.7) repeatedly with $l = n$ down to $l = j + 1$ to show inductively that $v_l^T D v_j = 0$ and as above, we conclude that v_j cannot be zero because it is a nontrivial linear combination of linearly independent vectors, $w_j, \ldots, w_n$. Further, $v_j \neq 0$ implies that $v_j^T D v_j > 0$ and this is important because the construction of all the v's with a smaller subscript (cf. Eq. (4.6)) involves division by $v_j^T D v_j$.

Thus we have shown recursively that v_j is well defined once v_{j+1} has been defined and that $v_l^T D v_j = 0$ for $l > j$. Because this relation is symmetric ($v_l^T D v_j = v_j^T D v_l$), we have in fact proven that the orthogonality holds for all $l \neq j$.

Remark: The WG–S algorithm was defined in a backward recursive form because the result is to be used to construct an upper triangular matrix factorization. If a lower triangular matrix factorization were intended, one would define the orthogonal vectors $\{v_j\}$ using a forward recursion.

The modified weighted Gram–Schmidt (MWG–S), described in Theorem VI.4.1, is a modest algebraic rearrangement of Eqs. (4.5) and (4.6) that has improved numeric characteristics and that is better suited for

[†]By *nontrivial* we mean that not all the w coefficients are zero.

FORTRAN implementation (cf. Appendix VI.A and Björck (1967)). Björck thoroughly analyzed the MWG–S algorithm and showed that it is far superior to the classical WG–S algorithm. Vectors computed using the classical WG–S procedure are not generally orthogonal and it is therefore necessary to iterate the weighted orthogonalization process. Further, to increase numeric accuracy, pivot strategies are introduced which involve ordering the vectors to avoid computing with nearly dependent vectors. Unlike the classical procedure, the modified algorithm produces almost orthogonal vectors, and reorthogonalization and pivot strategies are unnecessary. Jordan (1968) has carried out numeric experiments that corroborate Björck's analyses and has not only shown that the modified algorithm is numerically superior to the classical algorithm, but that it has accuracy comparable with the Householder orthogonal triangularization algorithm.

Theorem VI.4.1 (MWG–S Orthogonalization and Matrix Factorization) Let $w_1, \ldots, w_n$ be linearly independent N vectors with $N \geqslant n$ and let $D = \mathrm{Diag}(D_1, \ldots, D_N) > 0$. Consider the recursions

$$v_k^{(n)} = w_k, \qquad k = 1, \ldots, n \tag{4.8}$$

For $j = n-1, \ldots, 1$, cycle through equations (4.9)–(4.11):

$$\bar{D}_{j+1} = \left(v_{j+1}^{(j+1)}\right)^{\mathrm{T}} D v_{j+1}^{(j+1)} \tag{4.9}$$

$$\bar{U}(k, j+1) = \left(v_k^{(j+1)}\right)^{\mathrm{T}} D v_{j+1}^{(j+1)} / \bar{D}_{j+1}, \qquad k = 1, \ldots, j \tag{4.10}$$

$$v_k^{(j)} = v_k^{(j+1)} - \bar{U}(k, j+1) v_{j+1}^{(j+1)}, \qquad k = 1, \ldots, j \tag{4.11}$$

and

$$\bar{D}_1 = \left(v_1^{(1)}\right)^{\mathrm{T}} D v_1^{(1)} \tag{4.12}$$

Then $v_j^{(j)} = v_j$ for $j = 1, \ldots, n$, where v_j are the weighted orthogonal vectors given by the WG–S lemma, and

$$\begin{bmatrix} w_1^{\mathrm{T}} \\ \vdots \\ w_n^{\mathrm{T}} \end{bmatrix} D[w_1, \ldots, w_n] = \bar{U}\bar{D}\bar{U}^{\mathrm{T}} \tag{4.13}$$

where $\bar{U}$ is unit upper triangular.

Remark: Before proving this theorem we note that the inclusion of superscripts on the v's makes the algorithm appear, at first glance, to be rather complicated. In practice there is no need to include v superscripts and successive v's are written over the previous w values. The FORTRAN mechanization given in Appendix VI.A shows that the algorithm is reasonably compact and requires little more storage than that required for the W matrix.

Proof: For $j = n$ the result $v_n^{(n)} = v_n$ is trivial because both are equal to w_n. For $j = n - 1$ we have

$$v_k^{(n-1)} = v_k^{(n)} - \left(\left(v_k^{(n)}\right)^{\mathrm{T}} Dv_n/\bar{D}_n\right)v_n, \qquad k = 1, \ldots, n-1 \qquad \text{(cf. Eqs. (4.10) and (4.11))}$$

$$= w_k - \left(w_k^{\mathrm{T}} Dv_n/\bar{D}_n\right)v_n, \qquad k = 1, \ldots, n-1 \qquad \text{(cf. Eq. (4.8))} \tag{4.14}$$

Thus Eqs. (4.14) show that

$$\left(v_k^{(n-1)}\right)^{\mathrm{T}} Dv_n = 0, \qquad k = 1, \ldots, n-1 \tag{4.15}$$

and therefore $v_{n-1}^{(n-1)} = v_{n-1}$. The general result $v_j^{(j)} = v_j$ with $j \leqslant n-1$ now follows inductively; for suppose that

$$v_k^{(j)} = w_k - \sum_{\alpha=j+1}^{n} \left(w_k^{\mathrm{T}} Dv_\alpha/\bar{D}_\alpha\right)v_\alpha, \qquad k = 1, \ldots, j+1 \qquad \text{(cf. Eq. (4.6))} \tag{4.16}$$

This inductive hypothesis, together with Eqs. (4.10) and (4.11), gives

$$v_k^{(j-1)} = v_k^{(j)} - \left(\left(v_k^{(j)}\right)^{\mathrm{T}} Dv_j/\bar{D}_j\right)v_j$$

$$= w_k - \sum_{\alpha=j+1}^{n} \left(w_k^{\mathrm{T}} Dv_\alpha/\bar{D}_\alpha\right)v_\alpha - \left(\left(v_k^{(j)}\right)^{\mathrm{T}} Dv_j/\bar{D}_j\right)v_j \tag{4.17}$$

Because $v_\alpha^{\mathrm{T}} Dv_j = 0$ for $\alpha \neq j$, it follows from (4.16) that

$$\left(v_k^{(j)}\right)^{\mathrm{T}} Dv_j = w_k^{\mathrm{T}} Dv_j \tag{4.18}$$

Inserting (4.18) into (4.17) it follows that (4.16) holds with j replaced by $j - 1$, and the induction is complete.

The key result of this theorem, Eq. (4.13), is proven by combining the v–w defining relation, Eq. (4.6), with Eqs. (4.10) and (4.18), which together give

$$\bar{U}(k, j) = w_k^{\mathrm{T}} D v_j / \bar{D}_j \tag{4.19}$$

Thus

$$w_j = v_j + \sum_{k=j+1}^{n} \left(w_j^{\mathrm{T}} D v_k / \bar{D}_k \right) v_k \qquad \text{(cf. Eq. (4.6))}$$

$$= v_j + \sum_{k=j+1}^{n} \bar{U}(j, k) v_k \qquad \text{(cf. Eq. (4.19))}$$

Since this result holds for $j = 1, \ldots, n - 1$, one can write

$$\begin{bmatrix} w_1^{\mathrm{T}} \\ \vdots \\ w_n^{\mathrm{T}} \end{bmatrix} = \bar{U} \begin{bmatrix} v_1^{\mathrm{T}} \\ \vdots \\ v_n^{\mathrm{T}} \end{bmatrix} \tag{4.20}$$

where, by definition, $\bar{U}(j, j) = 1$, and $\bar{U}(i, j) = 0$ for $j = 1, \ldots, n$ and $i > j$. Result (4.13) now follows because

$$\begin{bmatrix} w_1^{\mathrm{T}} \\ \vdots \\ w_n^{\mathrm{T}} \end{bmatrix} D[w_1, \ldots, w_n] = \bar{U} \begin{bmatrix} v_1^{\mathrm{T}} \\ \vdots \\ v_n^{\mathrm{T}} \end{bmatrix} D[v_1, \ldots, v_n]\bar{U}^{\mathrm{T}} = \bar{U}\bar{D}\bar{U}^{\mathrm{T}}$$

since the v's are weighted orthogonal with

$$v_i^{\mathrm{T}} D v_j = \bar{D}(j) \delta_{i,j}$$

VI.5 Duality Relationships between Information and Covariance Algorithms

We have developed covariance-related and information-related algorithms as independent approaches to the linear discrete estimation problem. These different looking kinds of algorithms were developed as a result of seemingly unrelated criteria, but were demonstrated to be algebraically equivalent. The main reason for having independent developments is our belief that analysts are better equipped to understand, interpret, and extend these algorithms when they are presented as independent entities. Kaminski (cf. Kaminski *et al.* (1971)) has, however, pointed out that there are duality relations between covariance-related and information-related algorithms; and thus algorithms from one or the other class can be obtained using simple substitutions.

The duality relations that Kaminski recommends for square root algorithms are given in Table 1. The principle of Kaminski's observed duality is that time updating of covariance-related algorithms can be identified with measurement updating of information-related algorithms. We illustrate how the duality association of Table 1 is used by constructing a square root covariance time updating algorithm from the SRIF algorithm (V.2.8), which we write, in modified form, as

$$T\begin{bmatrix} \tilde{R} \\ R_\nu A \end{bmatrix} = \begin{bmatrix} \hat{R} \\ 0 \end{bmatrix}\begin{matrix} \}n \\ \}m \end{matrix} \tag{5.1}$$

The duality associations of the table mean that $\tilde{R}$, A, R_ν, and $\hat{R}$ are to be

Table 1

Duality Association between Information and Covariance Square Root Algorithms

Time update	Measurement update
$\tilde{S}_{j+1}$	$\hat{R}_j^{\mathrm{T}}$
$\Phi_j \hat{S}_j$	$\tilde{R}_j^{\mathrm{T}}$
$G_j(n, N_w)$	$A_j^{\mathrm{T}}(n, m)$
$R_w^{-1}(j)$	$R_\nu^{\mathrm{T}}(j)$[a]
$Q_j = R_w(j)^{-1} R_w(j)^{-\mathrm{T}}$	$R_\nu^{\mathrm{T}}(j) R_\nu(j)$[a]

[a] Our information algorithm formulations have assumed that the measurement errors are normalized, i.e., $z = Ax + \nu$ with $E(\nu\nu^{\mathrm{T}}) = R_\nu^{-1} R_\nu^{-\mathrm{T}}$ has been converted to normalized form by premultiplying $[A\ z]$ by R_ν; $R_\nu z = R_\nu A x + R_\nu \nu$ with $E[(R_\nu \nu)(R_\nu \nu)^{\mathrm{T}}] = I$. Thus to translate our information algorithms to the form required for the duality identification one merely replaces z and A by $R_\nu z$ and $R_\nu A$.

replaced by $(\Phi\hat{S})^{\mathrm{T}}$, G^{T}, $R_w^{-\mathrm{T}}$, and $\tilde{S}^{\mathrm{T}}$, respectively, and thus

$$T\begin{bmatrix} (\Phi\hat{S})^{\mathrm{T}} \\ \hline (GR_w^{-1})^{\mathrm{T}} \end{bmatrix} = \begin{bmatrix} \tilde{S}^{\mathrm{T}} \\ \hline 0 \end{bmatrix} \begin{matrix} \}\, n \\ \}\, N_w \end{matrix} \tag{5.2}$$

where as usual T is an orthogonal transformation.

This result can be verified by factoring Eq. (5.3):

$$\tilde{S}\tilde{S}^{\mathrm{T}} = \tilde{P} = \Phi\hat{P}\Phi^{\mathrm{T}} + GQG^{\mathrm{T}} = \begin{bmatrix} \Phi\hat{S} & GR_w^{-1} \end{bmatrix} \begin{bmatrix} (\Phi\hat{S})^{\mathrm{T}} \\ (GR_w^{-1})^{\mathrm{T}} \end{bmatrix} \tag{5.3}$$

Since the two factorizations of (5.3) can only differ by an orthogonal transformation multiplier, we have

$$\begin{bmatrix} \tilde{S} & | & 0 \end{bmatrix} = \begin{bmatrix} \Phi\hat{S} & | & GR_w^{-1} \end{bmatrix} T' \tag{5.4}$$

where T' is an orthogonal transformation. By applying the matrix transposition operation to (5.2) it can be seen that one could take $T' = T^{\mathrm{T}}$.

Remark: T can be chosen so that S is an upper triangular matrix, and in this form the result is compatible with the upper triangular measurement updating algorithm that was introduced in Chapter V. Note that one can construct such a triangularization using the elementary transformations discussed in Chapter IV and the construction is analogous to the one used to arrive at the matrix triangularization theorem.

Morf and Kailath [3] have constructed other duality associations and these promise to be quite useful for generating covariance-related and information-related dual algorithms.

APPENDIX VI.A FORTRAN Mechanization of the MWG–S Algorithm For Time Updating of the U–D Factors

In this appendix we FORTRAN mechanize the equations of the MWG–S algorithm of Section VI.4 and arrange the mechanization in a form that is consistent with the U–D measurement update mechanization of Appendix V.A.

Input x n vector state estimate
U Upper triangular matrix with $D(i)$ stored on the diagonals, this matrix corresponds to U and D
Φ State transition matrix
Q Diagonal process noise matrix $Q = \text{Diag}(Q(1), \ldots, Q(N_w))$
G $G(n, N_w)$ matrix involved with process noise (cf. Eq. (1.1))

Output x Time-updated state estimate $x := \Phi x$
U Time-updated upper triangular matrix factor with $D(i)$ stored on the diagonal

Remarks: (1) All of U is utilized in this algorithm mechanization, even though the input and output involve only the upper triangular portion of U. In addition, vectors $a(n)$, $v(n)$, and $D(n + N_w)$ are employed in the computations, with $D(n + l) = Q(l)$, $l = 1, \ldots, N_w$.

(2) If Φ has special structure, such as block triangularity, this should be exploited (cf. Section VII.4).

```
   N = n + N_w
   v(1) = 0.
   DO 20 J = n, 2, - 1
   DO 5 I = 1, J
5  D(I) = U(I, J)              @ D(1), . . . , D(J - 1) here used for
                                 temporary storage
   DO 10 I = 1, n
   U(I, J) = Φ(I, J)
```

Comment: The lower triangular portion of the input U is implicitly assumed zero.

```
   DO 10 K = 1, J - 1
10 U(I, J) = U(I, J) + Φ(I, K) * D(K)
   v(J) = 0.
   DO 15 K = 1, n
15 v(J) = v(J) + Φ(J, K) * x(K)
20 v(1) = v(1) + Φ(1, K) * x(K)
   D(1) = U(1, 1)
```

Comment: $v = \Phi x$, $U = \Phi U$, and diagonals retrieved.

```
   DO 30 J = 1, n
   x(J) = v(J)                 @ x = Φx
30 U(J, 1) = Φ(J, 1)           @ U = ΦU completed
```

Comment: It is assumed that $G(n, N_w)$ is stored in columns $n + 1$,

. . . , $n + N_w$ of U, the process noise is zero mean, and the N_w elements of Q are stored in $D(n + 1), \ldots, D(n + N_w)$.

```
      DO 100 J = n, 1, - 1
      σ = 0.
      DO 40 K = 1, N
      v(K) = U(J, K)                      @ U(J, .) = v^J of Eqs.
                                            (4.9)–(4.11)
      a(K) = D(K) * v(K)
40    σ = σ + v(K) * a(K)                 @ Eq. (4.9)
      U(J, J) = σ
      IF (J = 1) GO TO 100
      DINV = 1./σ
      JM1 = J - 1
      DO 60 K = 1, JM1
      σ = 0.
      DO 50 I = 1,N
50    σ = σ + U(K, I) * a(I)
      U(J, K) = σ                         @ Eq. (4.10), Ū(K, J) stored
                                            in U(J, K)
      DO 60 I = 1, N
60    U(K, I) = U(K, I) - σ * V(I)        @ Eq. (4.11)
100   CONTINUE
```

Comment: The 100 loop is the MWG–S algorithm Eqs. (4.9)–(4.12) with the time-updated values stored in U transpose.

```
      DO 110 J = 2, n
      DO 110 I = 1, J - 1
      U(I, J) = U(J, I)
      U(J, I) = 0.                        @ one need not explicitly zero
                                            the lower triangular part of U
110   CONTINUE
```

In Section VII.4 we discuss algorithms for the special but important cases where some of the variables are biases and others are correlated process noise.

References

[1] Bierman, G. J., Sequential square-root filtering and smoothing of discrete linear systems, *Automatica* **10**, 147–158 (1974).

This paper is primarily concerned with the smoothing problem, but it contains an expository description of the SRIF that is equivalent to the one presented here.

[2] Dyer, P., and McReynolds, S., Extension of square root filtering to include process noise, *J. Opt. Theory, Appl.* **3**, No. 6, 444–459 (1969).

In this paper the technique of including process noise was introduced, and the equivalence of the SRIF and Kalman filters was demonstrated. The paper is not written for the novice. It contains much that is of value, but many of the contributions are relegated to casual comments.

[3] Morf, M., and Kailath, T., Square root algorithms for least squares estimation, *IEEE Trans. Automatic Contr.* **AC-20**, 483–497 (1974).

The approach to square root estimation that is presented in this reference is based on Gram–Schmidt random variable orthogonalization procedures. The techniques presented in this paper differ from ours, but they lead to a number of interesting results. The geometric interpretations and the duality relations which are described in this paper are of special interest.

VII

Treatment of Biases and Correlated Process Noise

VII.1 Introduction

In this chapter we elaborate on the filter results of Chapter VI. In particular, we will be interested in applying the filter formulae to state vectors that are composed partly of biases and partly of exponentially correlated process noise. The reason for distinguishing between these special kinds of variables is that navigation problems frequently involve large numbers of bias parameters (viz., station location errors, ephemeris corrections, and planetary harmonics), and many random phenomena can be described or approximated by correlated process noise (viz., solar pressure and attitude control leaks).

By partitioning and expanding the filter formulae of Chapter VI, one can exploit the special character of these variables to obtain equation formulations that are computationally efficient and that reduce computer storage requirements. An additional attribute of the partitioning we use is that it facilitates analysis of the effects of individual parameters.

Let us partition the state vector as follows:

p_j Correlated process noise that is representative of a rather large class of Markov processes (cf. Jazwinski [1]). Examples include attitude control forces, solar pressure, and measurement-dependent errors that are correlated.

x_j States that are time varying (cf. Eq. (1.1)) but do not explicitly depend on white process noise (e.g., position and velocity errors).

y Bias parameters (e.g., constant acceleration errors, planet harmonics, station location errors, or ephemeris corrections).

The dynamic model for this partitioned system (cf. (VI.1.1)) is

$$\begin{bmatrix} p \\ x \\ y \end{bmatrix}_{j+1} = \begin{bmatrix} M & 0 & 0 \\ V_p & V_x & V_y \\ 0 & 0 & I \end{bmatrix} \begin{bmatrix} p \\ x \\ y \end{bmatrix}_j + \begin{bmatrix} w_j \\ 0 \\ 0 \end{bmatrix} \tag{1.1}$$

where V_p, V_x, and V_y are transition matrix elements and M is assumed, for convenience, to be a diagonal matrix†

$$M = \mathrm{Diag}(m_1, \ldots, m_{N_p}) \tag{1.2}$$

These matrices are, in general, time dependent (see Appendix VII.A), but we omit this dependence. Indeed, we omit the j dependence whenever the context makes the intended meaning clear.

VII.2 SRIF Data Processing for a System Composed Partly of Biases and Partly of Correlated Process Noise

In this section we delve further into the SRIF data processing method. Our intent is to exploit those features that are best suited for parameter estimation. An example of this is the introduction of the sensitivity matrix, which indicates the influence of the bias parameters on the estimate.

Suppose that at time t_j we have an *a priori* information array of the form

$$\begin{array}{c} \begin{matrix} \overbrace{\quad}^{N_p} & \overbrace{\quad}^{N_x} & \overbrace{\quad}^{N_y} & \overbrace{\quad}^{1} \end{matrix} \\ \begin{bmatrix} \tilde{R}_p & \tilde{R}_{px} & \tilde{R}_{py} & \tilde{z}_p \\ \tilde{R}_{xp} & \tilde{R}_x & \tilde{R}_{xy} & \tilde{z}_x \\ 0 & 0 & \tilde{R}_y & \tilde{z}_y \end{bmatrix} \begin{matrix} \}N_p \\ \}N_x \\ \}N_y \end{matrix} \end{array} \tag{2.1}$$

which represents the following data equations:

$$\begin{bmatrix} \tilde{R}_p & \tilde{R}_{px} & \tilde{R}_{py} \\ \tilde{R}_{xp} & \tilde{R}_x & \tilde{R}_{xy} \\ 0 & 0 & \tilde{R}_y \end{bmatrix} \begin{bmatrix} p_j \\ x_j \\ y \end{bmatrix} = \begin{bmatrix} \tilde{z}_p \\ \tilde{z}_x \\ \tilde{z}_y \end{bmatrix} - \begin{bmatrix} \tilde{\nu}_p \\ \tilde{\nu}_x \\ \tilde{\nu}_y \end{bmatrix} \tag{2.2}$$

†Nondiagonal M matrices occur infrequently in navigation applications, but might arise, for example, when studying the random motion of the Earth's pole or the force exerted on the hull of a ship by water waves.

The subscripts on the $\tilde{R}$ and $\tilde{z}$ terms are meant to associate corresponding matrices and states (p, x, and y). Equation (2.2) represents the *a priori* information at time t_j.

Remarks: (1) The term $\tilde{R}_{xp}$ will generally be nonzero when it is the result of a mapping (cf. Eq. (VI.2.29)).

(2) The block of zeros in (2.2) is preserved by the SRIF algorithm, (VI.2.28) and (VI.2.29).

A data processing algorithm for the partitioned system can be obtained by substituting the *a priori* information array (2.1) and the new data

$$z = A_p p_j + A_x x_j + A_y y + \nu \tag{2.3}$$

into the previous SRIF algorithm, Eq. (VI. 2.28). The presence, however, of the block of zeros in the *a priori* array and in the updated square root information matrix, which is by construction an upper triangular matrix, suggests the possibility of arranging the equations so that this block of zeros never appears.† The desired result is derived in the following steps:

(a) Rearrange the matrix rows of (2.1) and (2.3) to form

$$\begin{bmatrix} \tilde{R}_p & \tilde{R}_{px} & \tilde{R}_{py} & \tilde{z}_p \\ \tilde{R}_{xp} & \tilde{R}_x & \tilde{R}_{xy} & \tilde{z}_x \\ A_p & A_x & A_y & z \\ 0 & 0 & \tilde{R}_y & \tilde{z}_y \end{bmatrix} \tag{2.4}$$

This step corresponds to permuting equations and obviously has no effect on the estimate, covariance, or the least squares performance functional (VI.2.25).

(b) Apply a sequence of elementary Householder transformations (cf. Chapter IV) to partially triangularize (2.4), nulling the terms below the diagonal of the first $N_p + N_x$ columns. If $\hat{T}$ represents the product of the elementary transformations, then one may write

$$\hat{T}\begin{bmatrix} \tilde{R}_p & \tilde{R}_{px} & \tilde{R}_{py} & \tilde{z}_p \\ \tilde{R}_{xp} & \tilde{R}_x & \tilde{R}_{xy} & \tilde{z}_x \\ A_p & A_x & A_y & z \\ 0 & 0 & \tilde{R}_y & \tilde{z}_y \end{bmatrix} = \begin{bmatrix} \hat{R}_p & \hat{R}_{px} & \hat{R}_{py} & \hat{z}_p \\ 0 & \hat{R}_x & \hat{R}_{xy} & \hat{z}_x \\ 0 & 0 & \hat{A}_y & \hat{z} \\ 0 & 0 & \tilde{R}_y & \tilde{z}_y \end{bmatrix} \tag{2.5}$$

†Eliminating this block of superfluous storage reduces computer costs.

Note that the bottom matrix row of (2.5) is unchanged (because of Property IV.3 and the matrix triangularization theorem of Chapter IV) and that $\hat{T}$ is independent of the *a priori* on y, because

$$\hat{T} = \begin{bmatrix} \hat{T}_{px} & 0 \\ 0 & I_{N_y} \end{bmatrix}$$

and $\hat{T}_{px}$ is independent of y. As a result of these observations (2.5) can be modified to read

$$\hat{T}_{px} \begin{bmatrix} \tilde{R}_p & \tilde{R}_{px} & \tilde{R}_{py} & \tilde{z}_p \\ \tilde{R}_{xp} & \tilde{R}_x & \tilde{R}_{xy} & \tilde{z}_x \\ A_p & A_x & A_y & z \end{bmatrix} = \begin{bmatrix} \hat{R}_p & \hat{R}_{px} & \hat{R}_{py} & \hat{z}_p \\ 0 & \hat{R}_x & \hat{R}_{xy} & \hat{z}_x \\ 0 & 0 & \hat{A}_y & \hat{z} \end{bmatrix} \tag{2.6}$$

(c) Apply an orthogonal transformation to combine the y *a priori* with the modified y observation $\hat{A}_y y = \hat{z} - \hat{\nu}_y$:

$$\hat{T}_y \begin{bmatrix} \tilde{R}_y & \tilde{z}_y \\ \hat{A}_y & \hat{z} \end{bmatrix} = \begin{bmatrix} \hat{R}_y & \hat{z}_y \\ 0 & e \end{bmatrix} \tag{2.7}$$

Equations (2.6) and (2.7) are equivalent to the original partitioned version of (VI.2.28); the present algorithm, however, omits the block of $(N_p + N_x) \times N_y$ zeros present in the original problem. As will be explained in the next section, $\tilde{R}_y$ will always be an upper triangular matrix; and because of this the triangular algorithm described in Table VB.7 can be applied (cf. Appendix VII.B).

Let us turn our attention to the right side of Eqs. (2.6) and (2.7), and write the corresponding data equations for the x and y variables:

$$\hat{R}_x x_j + \hat{R}_{xy} y = \hat{z}_x - \hat{\nu}_x, \qquad \hat{R}_y y = \hat{z}_y - \hat{\nu}_y \tag{2.8}$$

Recalling that the estimate $\hat{x}_j$ and $\hat{y}$ satisfy (2.8) with the ν terms set to zero (cf. Chapter V), we have

$$\hat{y} = \hat{R}_y^{-1} \hat{z}_y \tag{2.9}$$

$$\hat{x}_j = \hat{R}_x^{-1} \hat{z}_x - \hat{R}_x^{-1} \hat{R}_{xy} \hat{y} \tag{2.10}$$

If one defines

$$\hat{x}_j^{\mathrm{c}} = \hat{R}_x^{-1}\hat{z}_x, \qquad \hat{S} = -\hat{R}_x^{-1}\hat{R}_{xy} \tag{2.11}$$

then (2.10) can be rewritten as

$$\hat{x}_j = \hat{x}_j^{\mathrm{c}} + \hat{S}\hat{y} \tag{2.12}$$

The significance of representation (2.12) is that x^{c}, called the *computed estimate*, is the estimate that would result if y had been omitted from the problem; and $\hat{S}$ is called the *sensitivity matrix* because

$$\hat{S} = \partial(x - \hat{x}^{\mathrm{c}})/\partial y = \partial(x - \hat{x})/\partial(y - \hat{y}) \tag{2.13}$$

If the diagonal elements of the y covariance matrix are $\sigma_y^2(1), \ldots, \sigma_y^2(N_y)$ then the matrix

$$\hat{P} = \hat{S}\,\mathrm{Diag}(\sigma_y(1), \ldots, \sigma_y(N_y)) = \left[\hat{P}_1, \ldots, \hat{P}_{N_y}\right] \tag{2.14}$$

is called the *perturbation matrix*. One can interpret the jth column of $\hat{P}$ as a "1σ" perturbation due to parameter y_j. There are two perturbation matrices one might speak about, depending on whether the y covariance used corresponds to the *a priori* before any data were included or corresponds to the current filter covariance. The first case is most frequently used and is motivated by observing that Eq. (2.8) can be written as

$$x = \hat{x}^{\mathrm{c}} + \sum_{j=1}^{N_y} \hat{S}_j y_j - \hat{R}_x^{-1}\hat{\nu}_x \tag{2.15}$$

where

$$\hat{S} = \left[\hat{S}_1, \ldots, \hat{S}_{N_y}\right]$$

Since the *a priori* $\tilde{\sigma}_y(j)$ bounds y_j, in a probabilistic sense, it follows that $\hat{P}_j = \hat{S}_j\sigma_y(j)$ bounds $\hat{S}_j y_j$. In an analogous manner one can write (cf. Eqs. (2.8) and (2.10))

$$x - \hat{x} = \sum_{j=1}^{N_y} \hat{S}_j(y_j - \hat{y}_j) - \hat{R}_x^{-1}\hat{\nu}_x \tag{2.16}$$

and infer that $\hat{S}_j\hat{\sigma}_y(j)$ bounds the error term $\hat{S}_j(y_j - \hat{y}_j)$. The perturbations $\hat{P}_j$ are only of qualitative significance when the y covariance is not diagonal, because when correlations exist between y_j and y_k it is incorrect to view $\hat{P}_j$ and $\hat{P}_k$ independently.

Remarks: (1) Note that $\hat{x}^c$ and $\hat{S}$ are independent of $\hat{R}_y$, so that these terms may be computed prior to deciding what the appropriate y *a priori* should be. This information is often used to decide which y parameters should be included in a given filter model.

(2) The matrix $\hat{P}^c$

$$\hat{P}^c = \hat{R}_x^{-1}\hat{R}_x^{-T} \tag{2.17}$$

is called the *computed covariance* and represents the covariance that would result if no y parameters were included in the problem.

(3) The matrix $\hat{P}_{\text{CON}}$

$$\hat{P}_{\text{CON}} = \hat{P}^c + \hat{S}\tilde{P}_y(0)\hat{S}^T \tag{2.18}$$

is a *consider covariance*, and

$$\hat{P}_{\text{CON}} = E\left[(x - \hat{x}^c)(x - \hat{x}^c)^T\right] \tag{2.19}$$

where $\hat{P}_{\text{CON}}$ is the error covariance that corresponds to the filter estimate $\hat{x}^c$ that was made without estimating the y parameters. In Chapter VIII we discuss computation of estimates that include some of the y parameters but which ignore the rest. We will also present error covariances to describe the effects of the unestimated parameters. Those results will generalize (2.11), (2.17), and (2.18).

(4) $\hat{x}^c$ and $\hat{S}$ are dependent on the p variables. However, our decision to arrange the state vectors with the p variables first resulted in representation (2.11), which does not explicitly involve p_j. As a result of this, formulae (2.8)–(2.19) are valid whether or not exponentially correlated p variables are present in the problem.

(5) The estimated vector x could be enlarged to include some of the y parameters and the results (computed estimates, computed error covariances, sensitivities, and consider covariances) would still hold. Of course the key terms of Eq. (2.8), ($\hat{R}_x$, $\hat{R}_{xy}$, $\hat{z}_x$, $\hat{R}_y$ and $\hat{z}_y$) have to be redefined. This is readily accomplished by letting $y = \begin{bmatrix} y_1 \\ y_2 \end{bmatrix}$ and expanding (2.8):

$$\begin{bmatrix} \hat{R}_x & \hat{R}_{xy_1} \\ 0 & \hat{R}_{y_1} \end{bmatrix} \begin{bmatrix} x_j \\ y_1 \end{bmatrix} + \begin{bmatrix} \hat{R}_{xy_2} \\ \hat{R}_{y_1 y_2} \end{bmatrix} y_2 = \begin{bmatrix} \hat{z}_x \\ \hat{z}_y \end{bmatrix} - \begin{bmatrix} \hat{\nu}_x \\ \hat{\nu}_{y_1} \end{bmatrix}$$

$$\hat{R}_{y_2} y_2 = \hat{z}_{y_2} - \hat{\nu}_{y_2} \tag{2.8a}$$

When (2.8a) and (2.8) are compared, the appropriate identification is clear.

(6) We mention in passing that one could compute estimates of successively larger dimension in a recursive fashion (cf. the triangular matrix inversion of Appendix IV.B).

(7) It is of course possible to extend the results of this section to include the colored noise p variables, i.e., to calculate sensitivities and consider covariances for the p variables. The extension is easily accomplished by replacing x by the augmented vector $\left[\begin{smallmatrix} p \\ x \end{smallmatrix}\right]$. This point is elaborated on in the remarks of Section VII.3.

(8) Expressions for the mapped (predicted) values $\tilde{x}^c$ and $\tilde{S}$ are much more complicated than (2.11) unless $\tilde{R}_{xp} = 0$; see, for example, Eqs. (3.9) and (3.11), which involve inversion of a larger matrix. In problems where mapped computed estimates or sensitivity matrices are of interest, one should first prepare the information array (2.1) by applying an orthogonal transformation to eliminate $\tilde{R}_{xp}$.

VII.3 SRIF Mapping Mechanization Including the Effects of Correlated Process Noise

One could arrive at the propagation algorithm to be presented in this section by substituting partitioned Φ and G matrices (cf. Eq. (1.1)) into the SRIF mapping algorithm, Eq. (VI.2.29), and rearranging the result. We choose not to proceed in this fashion because it is our opinion that an independent derivation of this important result will reinforce the basic ideas involved. Furthermore, the present derivation gives rise to a new algorithm,[†] one which can be applied to problems with a singular M matrix; where some of the m_i are zero, corresponding to white noise, is of particular importance.

Suppose that at time t_j we have an information array for p_j, x_j, and y, and we write it in data equation form[‡]

$$\begin{bmatrix} \hat{R}_p & \hat{R}_{px} & \hat{R}_{py} \\ \hat{R}_{xp} & \hat{R}_x & \hat{R}_{xy} \\ 0 & 0 & \hat{R}_y \end{bmatrix} \begin{bmatrix} p \\ x \\ y \end{bmatrix}_j = \begin{bmatrix} \hat{z}_p \\ \hat{z}_x \\ \hat{z}_y \end{bmatrix} - \begin{bmatrix} \hat{\nu}_p \\ \hat{\nu}_x \\ \hat{\nu}_y \end{bmatrix} \tag{3.1}$$

where the ν vector has zero mean and unity covariance. Our intent is to substitute the mapping equation (1.1) into this equation and arrive at an

[†]The reader may find it instructive to carry through the derivation of the propagation algorithm using the general result Eq. (VI.2.29) and compare this with (3.5).

[‡]$\hat{R}_{xp}$ is in most cases zero, because $\hat{R}$ generally comes about as the result of processing data, and is therefore triangular. When, however, process noise is involved, one might have time updates at points where there is no data, and in such cases the next time update would be initiated with a nontriangular information array.

equation of the same form for p_{j+1}, x_{j+1}, and y. From (1.1) we have

$$x_{j+1} = V_x x_j + V_p p_j + V_y y$$

which may be substituted into (3.1) to eliminate x_j, i.e.,

$$\begin{bmatrix} \hat{R}_p - (\hat{R}_{px} V_x^{-1}) V_p & (\hat{R}_{px} V_x^{-1}) & \hat{R}_{py} - (\hat{R}_{px} V_x^{-1}) V_y \\ \hat{R}_{xp} - (\hat{R}_x V_x^{-1}) V_p & (\hat{R}_x V_x^{-1}) & \hat{R}_{xy} - (\hat{R}_x V_x^{-1}) V_y \\ 0 & 0 & \hat{R}_y \end{bmatrix} \begin{bmatrix} p_j \\ x_{j+1} \\ y \end{bmatrix} = \begin{bmatrix} \hat{z}_p \\ \hat{z}_x \\ \hat{z}_y \end{bmatrix} - \begin{bmatrix} \hat{\nu}_p \\ \hat{\nu}_x \\ \hat{\nu}_y \end{bmatrix} \tag{3.2}$$

If this equation involved p_{j+1} instead of p_j, we would be done. In order to induce this state of affairs we focus on the mapping equation for p and combine it with the *a priori* data equation for w_j:

$$R_w w_j = z_w - \nu_w$$

and thus

$$R_w(p_{j+1} - M p_j) = z_w - \nu_w \tag{3.3}$$

The desired data equation is obtained by combining this equation with (3.2) and eliminating p_j. The elimination is carried out using an orthogonal transformation because it preserves the unit covariance characteristic of the error terms.

$$\tilde{T}_p \begin{bmatrix} -R_w M & R_w & 0 & 0 \\ \hat{R}_p - \overline{R}_{px} V_p & 0 & \overline{R}_{px} & \hat{R}_{py} - \overline{R}_{px} V_y \\ \hat{R}_{xp} - \overline{R}_x V_p & 0 & \overline{R}_x & \hat{R}_{xy} - \hat{R}_x V_y \end{bmatrix} \begin{bmatrix} p_j \\ p_{j+1} \\ x_{j+1} \\ y \end{bmatrix} = \tilde{T}_p \begin{bmatrix} z_w \\ \hat{z}_p \\ \hat{z}_x \end{bmatrix} - \tilde{T}_p \begin{bmatrix} \nu_w \\ \hat{\nu}_p \\ \hat{\nu}_x \end{bmatrix} \tag{3.4a}$$

where $\overline{R}_{px} = \hat{R}_{px} V_x^{-1}$ and $\overline{R}_x = \hat{R}_x V_x^{-1}$. The result is

$$\begin{bmatrix} R_p^* & R_{pp}^* & R_{px}^* & R_{py}^* \\ 0 & \tilde{R}_p & \tilde{R}_{px} & \tilde{R}_{py} \\ 0 & \tilde{R}_{xp} & \tilde{R}_x & \tilde{R}_{xy} \end{bmatrix} \begin{bmatrix} p_j \\ p_{j+1} \\ x_{j+1} \\ y \end{bmatrix} = \begin{bmatrix} z_p^* \\ \tilde{z}_p \\ \tilde{z}_x \end{bmatrix} - \begin{bmatrix} \nu_p^* \\ \tilde{\nu}_p \\ \tilde{\nu}_x \end{bmatrix} \tag{3.4b}$$

[†]

[†] A superscript symbol is needed to distinguish the top row partitioned entries of Eqs. (3.4b) from $\tilde{R}_p$, $\tilde{R}_{px}$, and $\tilde{R}_{py}$; and since this top row plays such a key role in smoothing (cf. Chapter X), we use the superscript *.

Because y is a bias it is unaffected by mapping. Thus,

$$\left[\hat{R}_y\ \hat{z}_y\right]_j = \left[\tilde{R}_y\ \tilde{z}_y\right]_{j+1} \tag{3.5}$$

i.e., the propagated y information array af time t_{j+1} is the same as it was at time t_j. Following our usual practice, we express the result in terms of the information array.†

Correlated Process Noise Mapping Algorithm

$$\tilde{T}_p \begin{bmatrix} \overbrace{-R_w M}^{N_p} & \overbrace{R_w}^{N_p} & \overbrace{0}^{N_x} & \overbrace{0}^{N_y} & \overbrace{z_w}^{1} \\ \hat{R}_p - \bar{R}_{px} V_p & 0 & \bar{R}_{px} & \hat{R}_{py} - \bar{R}_{px} V_y & \hat{z}_p \\ \hat{R}_{xp} - \bar{R}_x V_p & 0 & \bar{R}_x & \hat{R}_{xy} - \bar{R}_x V_y & \hat{z}_x \end{bmatrix} \begin{matrix} \}N_p \\ \}N_p \\ \}N_x \end{matrix}$$

$$= \begin{bmatrix} R_p^* & R_{pp}^* & R_{px}^* & R_{py}^* & z_p^* \\ 0 & \tilde{R}_p & \tilde{R}_{px} & \tilde{R}_{py} & \tilde{z}_p \\ 0 & \tilde{R}_{xp} & \tilde{R}_x & \tilde{R}_{xy} & \tilde{z}_x \end{bmatrix} \begin{matrix} \}N_p \\ \}N_p \\ \}N_x \end{matrix} \tag{3.6}$$

An observation to be made about Eqs. (2.6) and (2.7), for data inclusion, and Eqs. (3.5) and (3.6), for propagation, is that the algorithms involve two matrices

$$\begin{bmatrix} R_p & R_{px} & R_{py} & z_p \\ R_{xp} & R_x & R_{xy} & z_x \end{bmatrix} \tag{3.7}$$

and

$$[R_y \quad z_y] \tag{3.8}$$

To facilitate further discussion, the following notation is introduced:

$$S_p = \begin{bmatrix} R_p \\ R_{xp} \end{bmatrix}, \quad S_x = \begin{bmatrix} R_{px} \\ R_x \end{bmatrix}, \quad S_y = \begin{bmatrix} R_{py} \\ R_{xy} \end{bmatrix}, \quad S_z = \begin{bmatrix} z_p \\ z_x \end{bmatrix}$$

†The reason for preferring the information array is that this form is more compact and readily translates to a computer algorithm, e.g., Appendix VII.B.

and

$$S = [S_p \quad S_x \quad S_y \quad S_z]$$

Note that $[R_y \; z_y]$ is the y information array and that $[S_p \; S_x \; S_z]$ is the information array associated with estimating p and x if the y terms were not present; S_y is a kind of correlation matrix that relates the y estimates with p and x. Estimates and covariances of the p, x, and y variables are obtainable from S and $[R_y \; z_y]$, when desired, using the following equations:

$$\begin{bmatrix} p^c \\ x^c \end{bmatrix} = [S_p \quad S_x]^{-1} S_z \qquad \text{(computed estimate, independent of } y) \tag{3.9}$$

$$y_{\text{est}} = R_y^{-1} z_y \tag{3.10}$$

$$\text{Sen} = -[S_p \quad S_x]^{-1} S_y \qquad \text{(sensitivity)} \tag{3.11}$$

$$\begin{bmatrix} p \\ x \end{bmatrix}_{\text{est}} = \begin{bmatrix} p^c \\ x^c \end{bmatrix} + \text{Sen}\, y_{\text{est}} \qquad \text{(filter estimate)} \tag{3.12}$$

$$P^c = \begin{bmatrix} P_p^c & P_{px}^c \\ P_{xp}^c & P_x^c \end{bmatrix} = [S_p \quad S_x]^{-1} [S_p \quad S_x]^{-T}$$

$$(p\text{–}x \text{ computed covariance, independent of } y) \tag{3.13}$$

$$P_y = R_y^{-1} R_y^{-T} \qquad (y \text{ filter covariance}) \tag{3.14}$$

$$P = \begin{bmatrix} P_p & P_{px} \\ P_{xp} & P_x \end{bmatrix} = P^c + [\text{Sen}\, R_y^{-1}][\text{Sen}\, R_y^{-1}]^T$$

$$(p\text{–}x \text{ filter covariance}) \tag{3.15}$$

The "est" subscript on p, x, and y is to draw attention to the fact that these estimates could be either filter estimates with a ^, or predicted estimates, with a ~, and correspond to using S and $[R_y \; z_y]$ with either a ^ or a ~. (See Section VI.1 for a discussion of this notation.) Formulae (3.9)–(3.15) come about by partitioning the estimate–covariance relations of Section VII.2.

Miscellaneous Remarks: (1) The nomenclature introduced for x^c and p^c is somewhat restrictive. In Chapter IX we generalize the term "computed estimate" to mean an estimate that includes some but not all of the y parameters.

(2) The * terms appearing in Eq. (3.6) are of no interest for filtering purposes. They are, however, indispensible for smoothing and are discussed at length in Chapter X.

(3) The results for $\hat{x}$ given in Eqs. (3.9) and (3.11)–(3.13) reduce to the results given earlier, i.e., Eqs. (2.11), (2.12), and (2.17).

(4) The S propagation algorithm, Eq. (3.6), can be arranged so that no extra columns need be addended to the S matrix. An elegant computer mechanization is described in Appendix VII.B. Application-oriented analysts are advised to peruse the FORTRAN listings given in this appendix; it points up the simplicity of the SRIF mechanization.

(5) Partitioning can be used to good effect for the estimate–covariance Kalman filter equations; refer to Tables V.B.1 and V.B.2 and Eqs. (VI.3.1) and (VI.3.2). The results are obtained by routine substitution into these formulae; see, for example, Jazwinski [1].

VII.4 U–D Mapping Mechanization to Account for Bias Parameters and Correlated Process Noise

Many of the positive features of the SRIF carry over to the U–D covariance factorization because the SRIF R matrix and the U–D factors are related by $UD^{1/2} = R^{-1}I_{\pm}$, where $I_{\pm}$ is a diagonal matrix with entries of ± 1. In this section we present U–D results that are quite analogous to those for the SRIF. The U–D formulae simplify somewhat if the state vector is rearranged and x is placed before p, i.e., if (1.1) is rearranged as

$$\begin{bmatrix} x \\ p \\ y \end{bmatrix}_{j+1} = \begin{bmatrix} V_x & V_p & V_y \\ 0 & M & 0 \\ 0 & 0 & I \end{bmatrix} \begin{bmatrix} x \\ p \\ y \end{bmatrix}_j + \begin{bmatrix} 0 \\ w_j \\ 0 \end{bmatrix} \tag{1.1a}$$

The main reason for this permutation of variables is that the transition matrix is now almost triangular, and this simplifies some of the U–D mapping computations.

The U–D factors are partitioned consistently with the state vector and the symbol $\hat{}$ is used to indicate the values at time j, prior to applying the mapping (1.1a); $\tilde{}$ will be used to represent the values at time $j+1$, after the mapping

$$\begin{bmatrix} \hat{U}_x & \hat{U}_{xp} & \hat{U}_{xy} \\ 0 & \hat{U}_p & \hat{U}_{py} \\ 0 & 0 & \hat{U}_y \end{bmatrix}; \hat{D} = \mathrm{Diag}\left(\hat{D}_x, \hat{D}_p, \hat{D}_y\right) \quad (\textit{a priori} \text{ at time } j) \tag{4.1}$$

To better see how the U–D factors are mapped we decompose (1.1a) into two parts:

$$\begin{bmatrix} \bar{x} \\ \bar{p} \\ \bar{y} \end{bmatrix} = \begin{bmatrix} V_x & V_p & V_y \\ 0 & I & 0 \\ 0 & 0 & I \end{bmatrix} \begin{bmatrix} x \\ p \\ y \end{bmatrix}_j \tag{4.2}$$

$$\begin{bmatrix} x \\ p \\ y \end{bmatrix}_{y+1} = \begin{bmatrix} I & 0 & 0 \\ 0 & M & 0 \\ 0 & 0 & I \end{bmatrix} \begin{bmatrix} \bar{x} \\ \bar{p} \\ \bar{y} \end{bmatrix} + \begin{bmatrix} 0 \\ w_j \\ 0 \end{bmatrix} \tag{4.3}$$

Equation (4.2) is a formal artifice (i.e., $\bar{x} = x_{j+1}$, $\bar{p} = p_j$, and $\bar{y} = y$) and the U–D mapping corresponding to it corresponds to Eq. (3.2) that appeared in the SRIF development. Using $\bar{}$ values to indicate the U–D factors corresponding to the intermediate mapping (4.2) we find by direct expansion that

$$\left[\bar{U}_p \quad \bar{U}_{py} \quad \bar{U}_y \quad \bar{D}_p \quad \bar{D}_y\right] = \left[\hat{U}_p \quad \hat{U}_{py} \quad \hat{U}_y \quad \hat{D}_p \quad \hat{D}_y\right] \tag{4.4}$$

$$\bar{U}_{xp} = V_x\hat{U}_{xp} + V_p\hat{U}_p, \qquad \bar{U}_{xy} = V_x\hat{U}_{xy} + V_p\hat{U}_{py} + V_y\hat{U}_y \tag{4.5}$$

$$\bar{U}_x\bar{D}_x\bar{U}_x^{\mathrm{T}} = \left(V_x\hat{U}_x\right)\hat{D}_x\left(V_x\hat{U}_x\right)^{\mathrm{T}} \tag{4.6}$$

where $\bar{U}_x$ and $\bar{D}_x$ are computed using the modified Gram–Schmidt orthogonalization of Appendix VI.A.

Turning attention now to the second portion of mapping (4.3), the part involving the correlated process noise, we find by direct expansion of Eq. (VI.3.2) that

$$\tilde{U}_y = \bar{U}_y = \hat{U}_y, \qquad \tilde{D}_y = \bar{D}_y = \hat{D}_y \tag{4.7}$$

(i.e., U_y and D_y, the bias related terms, are unaffected by the mapping)

$$\tilde{U}_{py} = M\bar{U}_{py} = M\hat{U}_{py}, \qquad \tilde{U}_{xy} = \bar{U}_{xy} = V_x\hat{U}_x + V_p\hat{U}_{py} + V_y\hat{U}_y \tag{4.8}$$

$$\begin{bmatrix} \tilde{U}_x & \tilde{U}_{xp} \\ 0 & \tilde{U}_p \end{bmatrix} \begin{bmatrix} \tilde{D}_x & 0 \\ 0 & \tilde{D}_p \end{bmatrix} \begin{bmatrix} \tilde{U}_x^{\mathrm{T}} & 0 \\ \tilde{U}_{xp}^{\mathrm{T}} & \tilde{U}_p^{\mathrm{T}} \end{bmatrix}$$

$$= \begin{bmatrix} \bar{U}_x & \bar{U}_{xp} \\ 0 & \bar{U}_p \end{bmatrix} \begin{bmatrix} \bar{D}_x & 0 \\ 0 & \bar{D}_p \end{bmatrix} \begin{bmatrix} \bar{U}_x^{\mathrm{T}} & 0 \\ \bar{U}_{xp}^{\mathrm{T}} & \bar{U}_p^{\mathrm{T}} \end{bmatrix} + \begin{bmatrix} 0 & 0 \\ 0 & Q \end{bmatrix} \tag{4.9}$$

where $Q = E(w_j w_j^{\mathrm{T}})$ is diagonal. The best method for computing the ~ terms so that (4.9) is satisfied is at this time still unresolved, the reason being that there are two algorithms that appear worthy of recommendation.

To formulate the algorithms we delete the partition subscripts and write U as an upper triangular square matrix of dimension $N_x + N_p$ $(= N_{xp})$ and D a diagonal matrix with entries $D(1), \ldots, D(N_{xp})$.

Algorithm 1 (One-at-a-Time Processing) For $k = 1, \ldots, N_p$ cycle through Eqs. (4.10)–(4.13):

$$\tilde{D}(N_{xk}) = \bar{D}(N_{xk})M(k)^2 + Q(k) \tag{4.10}$$

$$\tilde{U}(N_{xk}, N_{xp} + j) = M(k)\bar{U}(N_{xk}, N_{xp} + j), \qquad j = 1, \ldots, N_y \tag{4.11}$$

$$\left.\begin{aligned} V(i) &:= \bar{U}(i, N_{xk}) \\ \tilde{U}(i, N_{xk}) &= \left(\bar{D}(N_{xk})M(k)/\tilde{D}(N_{xk})\right)V(i) \end{aligned}\right\} i = 1, \ldots, N_{xk} - 1 \tag{4.12}$$

where $N_{xk} = N_x + k$. Using the Agee–Turner algorithm (cf. Appendix III.C), compute the updated U–D factors of

$$\tilde{U}^{(k)}\tilde{D}^{(k)}(\tilde{U}^{(k)})^{\mathrm{T}} := \tilde{U}^{(k)}\tilde{D}^{(k)}(\tilde{U}^{(k)})^{\mathrm{T}} + c_k VV^{\mathrm{T}} \tag{4.13}$$

where in this equation $\tilde{U}^{(k)}$ and $\tilde{D}^{(k)}$ involve only the upper left $N_{xk} - 1$ rows and columns of the $\tilde{U}$ and $\tilde{D}$ matrices and

$$c_k = Q(k)\bar{D}(N_{xk})/\tilde{D}(N_{xk}) \tag{4.14}$$

Algorithm 2 (Modified Gram–Schmidt) Apply the modified Gram–Schmidt algorithm of Appendix VI.A to (4.15):

$$\tilde{U}\tilde{D}\tilde{U}^{\mathrm{T}} = W\,\mathrm{Diag}(D, Q)W^{\mathrm{T}} \tag{4.15}$$

where

$$W = \begin{array}{c} \begin{matrix} \overbrace{\quad}^{N_x} & \overbrace{\quad}^{N_p} & \overbrace{\quad}^{N_p} \end{matrix} \\ \left[\begin{matrix} \bar{U}_x & \bar{U}_{xp} & 0 \\ 0 & M\bar{U}_p & I \end{matrix}\right] \begin{matrix} \}N_x \\ \}N_p \end{matrix} \end{array}$$

Remark: As is often the case, the algorithm that appears most complicated (Algorithm 1 in this case) is in reality more economical in terms of computational efficiency and storage requirements. Unless N_p is large, however, the savings involved with using the first method do not appear to be significant.

One feature that is inherent to the use of triangular matrices is that *variable model* analyses are easily executed, i.e., one can compute estimates, estimate error covariances, sensitivities, and consider covariances that are based on different parameter models. These topics have already been discussed in Section VII.2 with regard to the SRIF. Using the analytic representation for the triangular matrix inverse (cf. Appendix IV.B) it is easy to show that if

$$U = \begin{bmatrix} U_x & U_{xy} \\ 0 & U_y \end{bmatrix}, \qquad D = \mathrm{Diag}(D_x, D_y)$$

then

$$\hat{S} = \partial(x - \hat{x})/\partial y = U_x U_y^{-1} \quad \text{(cf. Eq. (2.13))} \tag{4.16}$$

$$\hat{P}^{\mathrm{c}} = U_x D_x U_x^{\mathrm{T}} \quad \text{(cf. Eq. (2.17))} \tag{4.17}$$

Variable order estimates (cf. Eq. (2.12)) are computed a bit differently with the U–D factorization method than with the SRIF method. The SRIF method first computes $\hat{x}^{\mathrm{c}}$, the estimate based on a model without y parameters. Then $\hat{x}$, the optimal estimate based on the complete model, is obtained. The U–D method works in reverse, and starting with optimal estimates $\hat{x}$ and $\hat{y}$ it computes $\hat{x}^c$ via (2.12). The point of these comments is that the U–D filter algorithn shares the SRIF features that were described in Section VII.2.

VII.5 Computational Considerations

Arithmetic operation count comparisons of the various time propagation algorithms are not included. Thornton and Bierman [2] provide operation counts for time updating the U–D factors (discussed in Section VII.4), the triangular and nontriangular covariance square roots (corresponding to the Potter and Carlson measurement-updating algorithms), and the covariance matrix (using Eq. (VI.3.2)); the results are further elaborated upon in Thornton [3]. Operation counts for SRIF time updating are implicit in the tables of Chapter V (see also Bierman (1973b) and (1974)). There are unfortunately, a variety of factors that influence computational costs, and detailed comparisons involving all of these are, in our opinion, too unwieldy to be of significant value here. Instead we highlight points that are of major importance:

(1) All of the time propogation algorithms that we mentioned can take advantage of sparseness of the transition matrix. Significant computational savings are possible when the transition matrices are triangular (or block triangular) and triangular factorization such as the U–D and SRIF are used. State vectors that contain colored noise and bias parameters correspond to important special cases of block triangularity.

(2) Computational experience (restricted for the most part to orbit determination) has demonstrated that factorization algorithms are stable and accurate and, by comparison, the information matrix and covariance algorithms are not accurate. Comparing numerical results, even in apparently well-conditioned problems that have no obvious ill conditioning, has exposed order of magnitude discrepancies. Accuracy degradation is not always obvious, and including computational safeguards and numerical comparisons adds to the cost of the nonfactorization algorithms.

(3) Computer costs are sensitive to algorithm implementation and care should be taken so that efficient mechanizations are used. Indiscreet implementation can dramatically increase computational and/or storage requirements; in some cases the impression left is that a given algorithm is too expensive to warrant consideration. Early SRIF and Kalman filter algorithm mechanizations involved excessive amounts of storage and computation.

(4) When estimates and covariances are computed infrequently and the number of colored process noise terms is relatively small, the SRIF

propagation cost is competitive with the Kalman covariance propagation. The U–D propagation is also competitive with the Kalman covariance propagation. These comparisons depend on the form of the filter model; but they seem to apply to a spectrum of cases, including those common to orbit determination (cf. Thornton and Bierman [2] and Thornton [3]).

(5) Time propagation of covariance square roots using Householder orthogonal transformations is computationally more costly than the U–D algorithms of Section VII.4. Depending on the relative proportions of bias and colored noise variables that are included, the cost differential can be appreciable.

(6) For real-time filter applications involving minicomputers with limited computational precision the U–D factorization is superior to the Kalman filter mechanization because of its greater accuracy.

(7) An alternative U–D time propagation algorithm considered for some applications consists of creating the covariance matrix by direct multiplication, $(\Phi U)D(\Phi U)^{\mathrm{T}} + GQG^{\mathrm{T}}$, and then constructing the U–D factors via factorization (cf. Appendix III.A). This method is analogous to applying the Cholesky factorization to the normal matrix instead of using Householder orthogonal triangularization, and thus we do not recommend it. Criticisms of this scheme are that such methods are known to be a source of numerical error (cf. Section V.5) and the squaring up process involves more computation than does direct factor updating. Even where squaring up introduces computational economy, the savings generally do not warrent the threat of compromising numerical accuracy. We note further that in many ground-based applications filter computation comprises only a fraction of the total computational load; in such problems computational efficiency is not impacted by using factorization algorithms.

APPENDIX VII.A Exponentially Correlated Process Noise

A wide class of random processes can be described or approximated by first-order exponentially correlated process noise.† A formal mathematical description of such a process is

$$dp/dt = -(1/\tau)p + \omega \tag{A.1}$$

where τ is the time constant of the process and $\omega(t)$ is white, zero mean, process noise characterized by

$$E(\omega(t)) = 0, \qquad E(\omega(t)\omega(\tau)) = q_{\mathrm{con}}\,\delta(t-\tau) \qquad (\mathrm{A.2})^{\ddagger}$$

† First-order exponentially correlated noise is often called *colored noise*.

‡ $\delta(t)$ is the Dirac delta function.

where q_{con} is the noise density of the continuous time system. Vector-valued colored-noise problems are comprised of concatenations of scalar processes.

The variance corresponding to $p(t)$, $\sigma_p^2(t) = E(p^2(t))$, also satisfies a differential equation

$$d\sigma_p(t)/dt = -(2/\tau)\sigma_p^2(t) + q_{\text{con}} \tag{A.3}$$

Assuming that the process (A.1) has been operating indefinitely it is reasonable to assume that the statistics are in steady state $\sigma_p(t) = \sigma_p$ (steady state), so that

$$q_{\text{con}} = (2/\tau)\sigma_p^2 \tag{A.4}$$

When differential equations are used for filtering of continuous time systems q_{con} is generally defined via (A.4).

Remark: The term q_{con} given by (A.4) is a function of the time constant τ and will be large when a small time constant (near white noise) is used.

Since our applications involve p at discrete times t_j it is natural for us to convert Eq. (A.1) to a discrete form

$$p(t) = e^{-(t-\alpha)/\tau}p(\alpha) + \int_\alpha^t e^{-(t-\xi)/\tau}\omega(\xi)\,d\xi \tag{A.5}$$

It is easy to convert this result to recursive form, i.e.,

$$p_{j+1} = m_j p_j + w_j \tag{A.6}$$

where

$$p_j = p(t_j) \tag{A.7}$$

$$m_j = \exp(-(t_{j+1} - t_j)/\tau) \tag{A.8}$$

$$w_j = \int_{t_j}^{t_{j+1}} \exp(-(t_{j+1} - \xi)/\tau)\omega(\xi)\,d\xi \tag{A.9}$$

Equation (A.6) is the recursion used in most discrete applications.

Remarks: (1) $0 < m_j < 1$; the extremes occur for $\tau = 0$ and ∞, respectively.

(2) From (A.6) it follows that

$$E(p_{j+1}p_j) = m_j E(p_j^2) \tag{A.10}$$

(3) From (A.9) and (A.2) it follows that

$$E(w_j^2) = q_{\text{dis}} = 0.5\tau(1 - m_j^2)q_{\text{con}} \tag{A.11}$$

which shows how the discrete time noise variance q_{dis} is related to the continuous time noise variance q_{con}.

(4) The name "exponentially correlated process noise" is due to Eqs. (A.8) and (A.10).

(5) The case $m_j = 0$ corresponds to $\tau = 0$. In this case the p_j's form a white noise sequence.

In the continuous time case (A.4) is used to ascertain q_{con} and in the discrete time case we write

$$\sigma_p^2(j+1) = m_j^2\sigma_p^2(j) + q_{\text{dis}} \tag{A.12}$$

so that q_{dis} is obtained from the steady-state solution to (A.12)

$$q_{\text{dis}}^2 = (1 - m^2)\sigma_p^2 \tag{A.13}$$

Remark: One could use Eqs. (A.6), (A.12), and (A.13) to describe a discrete system that is piecewise constant. This is in fact the model used for navigation analyses at JPL.

More complicated correlated noise processes can be included within the framework of Eq. (A.6) if it is interpreted as a vector equation with the scalar m replaced by a matrix M. As an example, consider the second-order process

$$\ddot{p} + \lambda^2 p = \omega(t)$$

This process can be transformed to a discrete form with $p_1(j) = p(t_j)$ and $p_2(j) = \dot{p}(t_j)$; and the recursion takes the form

$$\begin{bmatrix} p_1 \\ p_2 \end{bmatrix}_{j+1} = M \begin{bmatrix} p_1 \\ p_2 \end{bmatrix}_j + \begin{bmatrix} w_1 \\ w_2 \end{bmatrix}_j$$

where $\Delta_j = t_{j+1} - t_j$ and

$$M = \begin{bmatrix} \cos(\lambda\,\Delta_j) & (1/\lambda)\sin(\lambda\,\Delta_j) \\ -\lambda\sin(\lambda\,\Delta_j) & \cos(\lambda\,\Delta_j) \end{bmatrix}$$

$$\begin{bmatrix} w_1 \\ w_2 \end{bmatrix}_j = (1/\lambda)\int_{t_j}^{t_{j+1}} \begin{bmatrix} (1/\lambda)\sin\lambda(t-\tau) \\ \cos\lambda(t-\tau) \end{bmatrix} \omega(\tau)\,d\tau$$

APPENDIX VII.B A FORTRAN Mechanization of the Epoch State SRIF

The SRIF algorithm that was described in this chapter is embodied in Eqs. (2.6) and (2.7) for data inclusion and (3.6) for propagation. FORTRAN mechanizations of these equations are presented here to demonstrate that computer mechanization of the SRIF is not complicated and does not require excessive amounts of storage. Equation (1.1) is the model our mechanization is based on, but instead of $x(t_j)$ given by

$$x(t_{j+1}) = V_x(t_{j+1}, t_j)x(t_j) + V_p(t_{j+1}, t_j)p_j + V_y(t_{j+1}, t_j)y \qquad \text{(B.1)}$$

we use x_j, where

$$x(t_j) = V_x(t_j, t_0)x_j + V_y(t_j, t_0)y \qquad \text{(B.2)}$$

and x_j is a pseudoinitial condition, one that compensates for the presence of the colored noise. A recursion for x_j is obtainable from (B.1), i.e.,

$$x_{j+1} = x_j + V_p(j+1)p_j \qquad \text{(B.3)}$$

where $V_p(j+1) = V_x^{-1}(t_{j+1}, 0)V_p(t_{j+1}, t_j)$. Note that there is no y term present because

$$V_y(t_{j+1}, 0) = V_x(t_{j+1}, t_j)V_y(t_j, 0) + V_y(t_{j+1}, t_j)$$

Equation (B.3) is used in our filter mechanization. By using x_j instead of $x(t_j)$, the transition matrix (cf. Eq. (1.1)) takes on a particularly simple form, i.e., $V_y = 0$ and $V_x = I$.

This choice of variables simplifies the filter propagation and also reduces computation in the smoother (to be discussed in Chapter X). In general, the penalty for using these transformed variables is that the measurement equations become more complicated. It happens that analysts working with classical orbit determination and space navigation problems often use essentially these variables in their formulations (i.e., they solve for the initial condition of the problem), and, as a result, the measurements need not be transformed since they are already expressed in terms of these variables. Such is the case with the orbit determination programs in use at the Jet Propulsion Laboratory.

Attributes of our SRIF mechanization are

A. Mapping of process noise one component at a time:
   1. Reduces computer storage requirements.
      Computer storage requirements are (essentially) independent of the number of process noise terms included in the model.
   2. Reduces computation.
      One-component-at-a-time processing requires only a partial matrix triangularization. Computational savings become more significant for problems involving repeated propagations with no

data present. Direct coding of the Householder algorithm (see the FORTRAN listings that follow) avoids storing the information array in a form that the general purpose computer library subroutines can accept, and reshuffling of the result for subsequent use. (Indeed, it is the author's experience that the computer code required to shuffle the array before and after using the general purpose JPL computer library routines utilizes nearly as many FORTRAN instructions as does the inline programming of the entire algorithm.) An additional benefit of inline code is that it is simpler and more efficient than are library routines. The reason for this is that computer library routines are designed for a variety of applications and as such they contain options and numerical enhancements that are not needed in our applications.

3. Auxiliary arrays for smoothing are generated.
   The smoother mechanization (Chapter X) is most efficiently obtained from the one-component-at-a-time processed filter arrays.

B. Storage of the bias parameter information array using vector storage: Many of the parameters occuring in space applications are constant. Our computer implementation exploits this observation to reduce computations and storage requirements. First, the block of zeros R_{yp} and R_{yx} is removed from the last row of Eq. (2.4), and this saves $N_y * (N_p + N_x)$ elements. A further reduction in storage results from the observation that the array R_y is always upper triangular. Storing this array as a vector, i.e.,

$$R_y = \begin{bmatrix} R_y(1) & R_y(2) & R_y(4) & \\ & R_y(3) & R_y(5) & \text{etc}. \\ & & R_y(6) & \end{bmatrix}$$

leads to a saving of $N_y * (N_y - 1)/2$ storage locations. To capitalize on this storage savings a Householder triangularization is included which utilizes vector stored inputs.

Benefits resulting from the reduction in storage requirements are:

1. Cost is reduced (as a result of smaller computer core-time products).
2. Our implementation allows for the inclusion of parameter vectors of large dimension. For example, filter problems with more than 200 parameters are solved on JPL's UNIVAC 1108 computer.

C. New mapping mechanization to accommodate white and near white correlated noise:[†]
Our propagation algorithm, Eq. (3.6), is an improvement over the original Dyer–McReynolds algorithm, Eq. (VI.2.29), because we require no special treatment of white and near white noise. Additional advantages of our computer mechanization are that the coding is simpler and computation is reduced (although the savings are negligible unless the filter has large dimension).

The state vector for the pseudoepoch state model (B.3) is taken to be

$$\begin{bmatrix} p_j \\ x_j \\ y \end{bmatrix} \begin{matrix} \}N_p \\ \}N_x \\ \}N_y \end{matrix}$$

and the notation of Section VI.3 is used, i.e., the SRIF array is stored in the two matrices S and R_y, where

$$S(N_{px}, N_{\text{tot}}) \equiv \begin{bmatrix} \overbrace{S_p}^{N_p} & \overbrace{S_x}^{N_x} & \overbrace{S_y}^{N_y} & \overbrace{S_z}^{1} \end{bmatrix}$$

and the last N_y entries of R_y contain the vector z_y. We write $N_{px} = N_p + N_x$, $N_{\text{tot}} = N_{px} + N_y + 1$, and $NR_y = N_y(N_y + 3)/2$ (the dimension of the augmented R_y matrix).

Our recursive algorithms for measurement updating and propagation in time are arranged so that new values replace old ones. Our reasons for reusing variables is that this technique avoids the often excessive use of unessential auxiliary storage, and it turns out that the computer code is more compact and efficient.

SRIF Measurement Processing (FORTRAN Code)

Assume that the matrix $[A \;\; z]$ corresponding to the m component measurement

$$z = A \begin{bmatrix} p \\ x \\ y \end{bmatrix} + \nu$$

[†]That is, correlated noise having small correlation times. In this case the augmented state transition matrix is poorly conditioned and near singular.

is stored in the lower portion of S, i.e.,

$$\left.\begin{aligned} S(N_{px}+I, J) &= A(I, J), \qquad J = 1, \ldots, N_{px}+N_y, \\ S(N_{px}+I, N_{\text{tot}}) &= z(I) \end{aligned}\right\} I = 1, \ldots, m$$

Computer code corresponding to

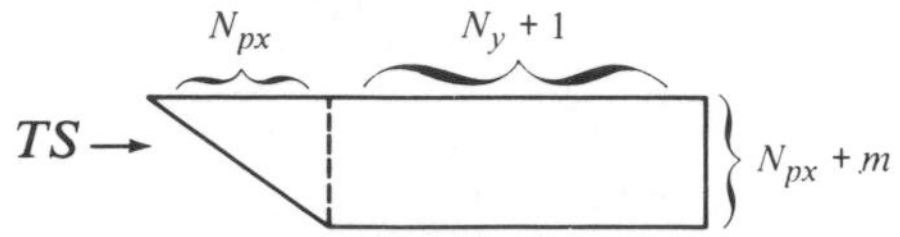

(B.4)

where T is orthogonal now follows:

Remark: For notational convenience we omit subscripts and write, for example, Npx and Ny instead of N_{px} and N_y.

Input Npx Number of stochastic and dynamic variables
Ny Number of bias parameters
m Number of measurements being processed
$S(L, N\text{tot})$ *A priori* information array, $L = Npx + m$, $N\text{tot} = Npx + Ny + 1$
Lower part of S contains measurements

Output
$S(L, N\text{tot})$ *A posteriori* information array, lower part of S contains measurement data to be combined with y *a priori*

```
    DO 4 J = 1, Npx
    σ = Z                          @ Z = zero
    DO 1 I = J, L
    v(I) = S(I, J)                 @ v(Npx) is a work vector
    S(I, J) = Z
1   σ = σ + v(I) ** 2
    IF (σ ≤ Z) GO TO 4
```

Comment: If $\sigma = Z$, column J is zero and this step of the reduction is omitted.

```
    σ = SQRT(σ)
    IF (v(J) > Z) σ = −σ
```

```
  S(J, J) = σ
  v(J) = v(J) - σ
  σ = ONE/(σ * v(J))              @ ONE = 1.
  DO 3 K = J + 1, Ntot
  α = Z
  DO 2 I = J, L
2 α = α + S(I, K) * v(I)
  α = α * σ
  DO 3 I = J, L
3 S(I, K) = S(I, K) + α * v(I)
4 CONTINUE
```

The bottom m rows of S, starting at column $N_{px} + 1$, are combined with R_y (stored as a vector)

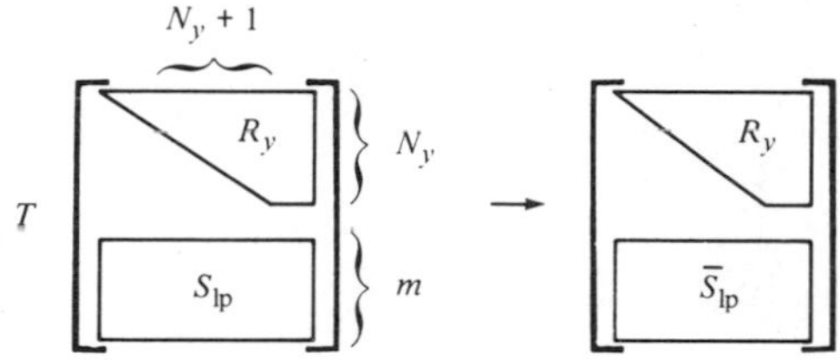

(B.5)

where T is orthogonal, S_{lp} is the lower right portion of S, and $\bar{S}_{lp}$ is theoretically zero (except for the last column). Computer code corresponding to (B.5) follows.

Remark: The main difference between (B.4) and (B.5) is that implementation of (B.5) utilizes Ry stored as a vector. We continue to omit time subscripts in our quasi-FORTRAN implementation.

Input

Ny	Number of bias parameters
m	Number of measurements being processed
$Ry(NRY)$	Triangular, vector stored, SRIF *a priori* y information array, $NRY = Ny(Ny + 3)/2$
$\bar{S}(m, Ny + 1)$	Comprised of the bottom m rows and the last $Ny + 1$ columns of S; this is the y portion of the observation

Output

$Ry(NRY)$	Triangular, vector stored, SRIF *a posteriori* y information array

```
   KK = 0
   DO 40 J = 1, Ny
   KK = KK + J                         @ KK = (K, K)
   σ = Z                               @ Z = zero
   DO 10 I = 1, m
10 σ = σ + S̄(I, J) ** 2                @ S̄(I, J) = S(Npx + I, Npx + J)
   IF (σ = Z) GO TO 40
```

Comment: If column J of observation is zero, this step of the reduction is omitted.

```
   σ = SQRT(σ + Ry(KK)) ** 2
   IF (Ry(KK) > Z) σ = −σ
   α = Ry(KK) − σ
   Ry(KK) = σ
   β = ONE/(σ * α)                     @ ONE = 1.
   JJ = KK
   L = J
   DO 30 K = J + 1, Ny1                @ Ny1 = Ny + 1
   JJ + JJ + L
   L = L + 1
   σ = α * Ry(JJ)
   DO 20 I = 1, m
20 σ = σ + S̄(I, J) * S̄(I, K)          @ S̄(I, K) = S(Npx + I, Npx + J)
   σ = σ * β
   Ry(JJ) = Ry(JJ) + σ * α
   DO 30 I = 1, m
30 S̄(I, K) = S̄(I, K) + σ * S̄(I, J)
40 CONTINUE
```

Observe that both segments of the computer code, corresponding to (B.4) and (B.5), are of interest. Equation (B.4) can be applied (with $Ny = 0$) to triangularize a system of $N_{px} + m$ equations in N_{px} variables; (B.5) can be used (with $N_{px} = 0$) for recursive triangularization of overdetermined systems where the results are stored in vector form (to economize on storage).

SRIF Propagation from T to T + DT (FORTRAN Code)

Input DT Propagation interval
Np Number of colored noise variables

Nx	Number of dynamic variables
Ny	Number of bias parameters
Nd	Number of dynamically related colored noise variables
$\tau(Np)$	Time constants corresponding to colored noise variables
$Vp(Nx, Nd)$	The first Nd columns of the VP matrix correspond to the dynamic parameters (cf. Eq. VII.B.3); the last $Np - Nd$ columns are omitted because they are in theory zero
$Rw(Np)$	Process noise standard deviation reciprocals
$S(Npx + Np, N\text{tot})$	The top Npx rows of S contain the SRIF array corresponding to the p and x variables; the bottom p rows are used to store smoothing-related terms

Output SIG(np)	Smoothing-related coefficients
$S(N_{px} + N_p, N\text{tot})$	Time-updated array with smoothing-related terms stored in the bottom portion of S

Comment: Parameters needed for the propagation step are $Npx = Np + Nx$, $N\text{tot} = Npx + Ny + 1$, and $DM(J) = \text{EXP}(-DT/\tau(J))$, $J = 1, \ldots, Np$. If $\tau(J) = Z$, $DM(J) = Z$, where Z = zero.

```
      DO 106 J1 = 1, Np
      IF (J1 > Nd) GO TO 102
      DO 101 I = 1, Npx
      DO 101 K = 1, Nx
101   S(I, 1) = S(I, 1) - S(I, Np + K) * Vp(K, J1)
```

Remark: Step 101 corresponds to Eq. (VII.3.2) with $Vx = I$ and $Vy = 0$.

```
102   α = - Rw(J1) * DM(J1)
      σ = α ** 2
      DO 103 I = 1, Npx
      v(I) = S(I, 1)                 @v(Npx) is a work vector
103   σ = σ + v(I) ** 2
      σ = SQRT(σ)
      α = α - σ
      SIG(J1) = σ                    @SMOOTH
      σ = ONE/(σ * α)                @ ONE = 1.
      DO 105 J2 = 2, Ntot
      δ = z
  c   IF (J2=Ntot) δ=α * Zw(J1)
```

Comment: Colored noise is generally zero mean; thus Zw, zero in most cases, is not explicitly included in the code.

```
      DO 104 I = 1, Npx
104   δ = δ + S(I, J2) * v(I)
      δ = δ * σ
      L = J2 - 1
      IF (J2 > Np) L = J2
      S(Npx + J1, L) = δ * α                          @SMOOTH
      DO 105 I = 1, Npx
105   S(I, L) = S(I, J2) + δ * v(I)
c     S(Npx+J1, Ntot)=S(Npx+J1, Ntot)+δ * Zw(J1)
                                                      @SMOOTH
      δ = α * Rw(1) * σ
      S(Npx + J1, Np) = Rw(J1) + δ * α                @SMOOTH
      DO 106 I = 1, Npx
106   S(I, Np) = δ * v(I)
```

Remark: Terms with @SMOOTH are used only for smoothing.

Figure VII.1 describes how the two key blocks, data processing and time propagation, are used in a filter mechanization.

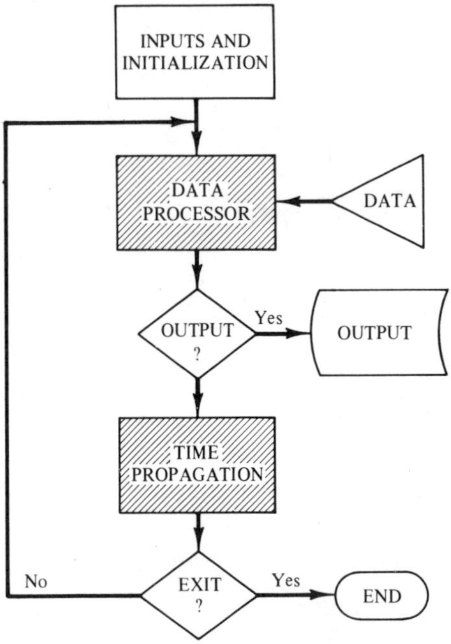

Fig. VII.1 Filter operation flow chart.

References

[1] Jazwinski, A. H., "Stochastic Processes and Filtering Theory." Academic Press, New York (1970).

This is a good reference. Chapters 7 and 8 contain many useful formulae and comments. They reflect the author's extensive filtering experience.

[2] Thornton, C. L., and Bierman, G. J., Gram–Schmidt algorithms for covariance propagation, *Proc.* 1975 *IEEE Conf. on Decision and Control, Houston, Texas*, 489–498.

Our *U–D* time updating of Section VII.4 is based on the developments of this paper. Computational efficiency of the *U–D* factorization (using modified Gram–Schmidt updating), the Potter and Carlson square root factorizations (using Householder orthogonal transformation updating), and the Kalman covariance time update are compared. Arithmetic operation counts are included and the effects of bias parameters and colored noise variables are distinguished.

[3] Thornton, C. L., "Triangular Covariance Factorizations for Kalman Filtering," Ph.D. dissertation, UCLA School of Engineering, Department of Systems Science (1976).

This dissertation goes into greater detail than [2] concerning computational efficiency. It includes a computer-simulated orbit determination case study that highlights the effects of numerical errors on the Kalman filter and demonstrates the efficiency and accuracy of the *U–D* filter.

VIII

Covariance Analysis of Effects Due to Mismodeled Variables and Incorrect Filter a Priori Statistics

VIII.1 Introduction

In the previous chapter we indicated how, using triangular factorization, one might analyze the effects of unestimated parameters. In this chapter and the next we expand on this notion and consider not only the effect that unestimated parameters have on filter accuracy but also the effects of applying a filter algorithm based on incorrect *a priori* covariance statistics.

The least squares algorithms described in Chapters II and V–VII employ weighting matrices that are designed to combine *a priori* information and new data in, what is believed to be, a "rational manner." It was pointed out in Chapter II that, when the weights are chosen to be *a priori* covariance statistics, the least squares estimate is also a minimum variance estimate. Thus by "rational manner" we mean that least squares weighting matrices are chosen that correspond to reasonable or likely *a priori* covariance statistics. An important feature of our estimation algorithms is that error covariances for the estimates are readily available or can be readily computed. This covariance knowledge is used to measure confidence in the estimate; and discrepancies between the estimate, the associated error covariances, and the measurement residual errors are generally cause for alarm.

Errors due to selecting incorrect *a priori* statistics or mismodeling of the problem can result in a deterioration of filter accuracy that sometimes results in *filter divergence*, i.e., the estimate error $x - \bar{x}$ increases or remains unacceptably large. It has been observed that divergence is often accompanied by overly optimistic filter covariances. An optimistic filter covariance is a matrix that is smaller† than the actual estimate error covariance. When the filter covariance is too small, the filter tends to reject data or weight it too lightly, and we say that the filter has learned the estimated state too well. Various strategies have been proposed to prevent the filter covariance from becoming too optimistic. One of these strategies, *consider filtering*, will be discussed in Section VIII.2.

A consider filter involves partitioning the model states into two sets X and y; X is to be estimated and y is to be considered. It is assumed that y is zero mean and the estimate of y will remain zero, independent of the measurement information. Rationalizations of this assumption are that the terms designated y cannot be accurately estimated, or estimating them might lead to an overly optimistic y filter covariance. A result of not estimating y is that it retains its *a priori* covariance P_y, and this covariance then appears as an error source to corrupt (increase) the filter covariance P_X.

Suppose that the *a priori* statistics are, as is generally the case, unknown and that values are assumed for use in the filtering algorithm. In this case the estimate, no longer optimal or minimum variance, may be degraded and the *computed covariance*‡ may not be reliable as a measure of confidence. It is, therefore, logical to cogitate about computing the *actual* or *true* covariance of the estimate and using this as a confidence gauge. Covariance performance of a filter estimate is generally obtained by hypothesizing a set of "actual" statistics and computing the estimate error covariance. Computation of an actual covariance is of qualitative value in the detection of covariance sensitivity to variations in the *a priori* statistics and is of quantitative value because it provides a statistical accuracy bound for the estimate based on an incorrect model. Performance analyses based on "actual" covariances are of prime importance both in determining engineering feasibility and in estimating the probability of a filter's success.

The idea of computing an actual covariance is readily generalized to include analyzing the effects of bias and colored noise parameters that have been omitted from the filter model. In principle the problem of omitted parameters is subsumed under the incorrect *a priori* problem; the filter assumes zero error variances for these parameters. The problem of unestimated parameters is, however, of special importance, and as a result

†If matrices P_1 and P_2 are positive definite (cf. Chapter III), then $P_1 < P_2$ means that $P_2 - P_1$ is positive definite; and we say that "P_1 is less than P_2". In like manner $P_1 > P_2$ is read "P_1 is greater than P_2," and it means that $P_1 - P_2$ is positive definite.

‡The term "computed covariance" here means the matrix thought to be the covariance of the estimate. It is based on the assumed statistics used in the filter algorithm computation, i.e., it is the P matrix used in the Kalman covariance filter and is $R_X^{-1}R_X^{-T}$ in the SRIF, where X represents the vector of all estimated parameters.

it receives a separate treatment. The "actual" covariance that arises in the evaluation of a filter model that omits certain parameters is called a *consider covariance*, and we speak of considering the effects of the unestimated parameters, called *consider parameters*.

Remark: Consider parameters appear in both consider covariance analysis and in consider filtering. There is no inconsistency here because in both cases these parameters are unestimated; the difference is that in the case of the consider covariance the filter model is oblivious to the presence of the consider parameters.

One is motivated to omit parameters from a filter model for various reasons; two of which are that it is cost effective to deal with a lower-dimensional simplified model (if accuracy is not too severely compromised), and sometimes large-dimensioned models are overly sensitive. The notion of a consider covariance came about as the result of a desire to evaluate covariance performance of lower-dimensional filter models that did not include all of the (bias) parameters present in the original problem. Because of a desire to distinguish it from the consider covariance, analysts in certain circles have come to label as a *generalized consider covariance* one that involves not only the effects of unestimated parameters, but also unestimated colored noise and incorrect *a priori* statistics for the filter algorithm. Thus the term "generalized consider covariance" is used in place of actual covariance.

Having spoken of consider filters and consider covariances we caution the reader to avoid the semantic pitfall of associating the consider covariance with the consider filter; they are different. A consider filter involves modifying the filter computed covariance to reflect the potential presence of unestimated variables. Most importantly, the filter estimate is modified in the process, either explicitly through a filter gain or implicitly through the filter parameters, e.g., R and z. Consider covariances, on the other hand, do not affect filter estimates and as a result are usually computed offline, either to aid in preliminary filter design or to aid in postflight data analysis.

To illustrate the basic ideas involved in the analysis of the effects of incorrect *a priori*, we review, in Section VIII.2, classical (constant parameter) batch filter consider analyses. The key result of that section, Eqs. (2.8) and (2.9), is a generalized consider covariance that includes the effects of incorrect *a priori* statistics and the effects of consider parameters.

In Section VIII.3 the generalized consider covariance for the Kalman filter is developed. The analysis includes construction of sensitivity vectors for the consider bias parameters.

The error analyses of Sections VIII.2 and VIII.3 do not include the effects of mismodeling the transition or measurement matrices. In our navigation applications the need has not arisen for a general mismodeling evaluation capability, such as that described by Sage and Melsa [1].† Our main concern with mismodeling occurs in relation to the selection of colored noise model time constants. Engineering judgement often dictates candidate time constants, and we rely on covariance performance analyses to detect model sensitivities due to these parameter choices. In Section VIII.4 we show that one can evaluate these effects by considering an augmented state problem with incorrect *a priori* statistics. The pragmatic importance of our treatment of the colored noise mismodeling problem lies in (a) the fact that no new software need be coded and checked out, and (b) the existing software is not burdened with the curse of generality, which so often punishes the average computer user with excessive storage and computation costs (even when he has no need of the full generality of the program he is operating).

VIII.2 Consider Analyses for the Batch (Constant Parameter) Filter

We begin our study of covariance error analysis with the batch filter because of its frequent use by navigation analysts and because our treatment provides a unified approach to a variety of special problems. In this section expressions are derived to evaluate the effects of the following incorrect *a priori* assumptions:

(a) an incorrect *a priori* covariance for the estimated parameters X,
(b) an incorrect *a priori* covariance for the measurement noise (in the filter it is assumed to have unit covariance, while in actuality it has covariance P_v^a),
(c)‡ an incorrect *a priori* covariance for the consider parameters y_c of the consider filter; including the problem of neglecting the y_c parameters in the estimation process, and
(d) unestimated parameters that have an *a priori* correlation with the estimated parameters.

Problems of types (a) and (b) arise quite naturally since all of the data processing algorithms we have discussed require *a priori* covariances for the state and measurements. The values assumed for these covariances

†The reader who needs to study the effects of general mismodeling of the transition and measurement matrices is advised to consult this reference. See also the aside on pages 188–189, where we indicate how such effects can be included within the framework of our data equation approach to error analysis.

‡The type (c) problem formulation allows us to evaluate the consider filter and the consider covariance using a unified approach.

may significantly degrade the estimate or may introduce unwanted sensitivity. Techniques to detect these phenomena are essential for the design of reliable filter software.

Problems of type (c) arise, for example, when one is estimating an augmented state vector $\begin{bmatrix} X \\ y \end{bmatrix}$, and the y covariance decreases too rapidly as new data are included. If no corrective action were taken the y covariance would approach zero, indicating that y was nearly perfectly known, and the augmented state filter would in effect reduce to an X state filter. Believing that an X state filter will incorrectly estimate X, it is sometimes proposed that y be included in the filter formulation, but that we agree not to estimate it, i.e., after each batch of measurements is processed, we set the y estimate back to its *a priori* zero value and retain the y covariance at its *a priori* value. This filter algorithm rational is called a *consider filter* because the filter uses information about the unestimated (consider) parameters.

Remark: An alternative method of preventing the y covariance from decreasing too rapidly is to model y as a colored noise. This method is often effective when y is known to be (slowly) varying.

In certain problems it seems either appropriate or expedient to deal with a lower-dimensional problem, and one ignores the y parameters in the filter. This case, too, is considered in case (c), and there we calculate the covariance error due to neglecting y in the estimator.

Problems of type (d) occur when estimated and unestimated parameters are dynamically related (e.g., an unestimated nongravitational acceleration bias error integrates into position and velocity errors, which are estimated). Another example of a type (d) problem occurs when data is processed in sequential batches. In this case the *a posteriori* estimate becomes correlated with the unestimated parameters.

The statistical characteristics of the estimation problem to be discussed in this section are summarized in the following equations:

$$\bar{X}, \quad \bar{y} = 0 \qquad \text{(filter estimates)} \tag{2.1}$$

$$\bar{P} = \begin{bmatrix} \bar{P}_X & \bar{P}_{Xy} \\ \bar{P}_{yX} & \bar{P}_y \end{bmatrix} \qquad \text{(augmented state filter covariance)} \tag{2.2}$$

$$\bar{P}^{\mathrm{a}} = \begin{bmatrix} \bar{P}^{\mathrm{a}}_X & \bar{P}^{\mathrm{a}}_{Xy} \\ \bar{P}^{\mathrm{a}}_{yX} & \bar{P}^{\mathrm{a}}_y \end{bmatrix} \qquad \text{(augmented state true covariance)} \tag{2.3}$$

The overbar symbol represents either *a priori* or *a posteriori* values (i.e. ~ or ^, cf. Section VI.1), and $\tilde{X}$ and $\hat{X}$ have computed covariances $\tilde{P}_X$ and $\hat{P}_X$, respectively. The latter covariance is assumed to come about as a result of applying an unbiased linear estimation algorithm to the *a priori* estimate and a perhaps incorrectly weighted measurement vector. The actual or true covariances $\tilde{P}_X^{\mathrm{a}}$ and $\hat{P}_X^{\mathrm{a}}$, respectively, correspond to the estimates $\tilde{X}$ and $\hat{X}$, with $\hat{P}_X^{\mathrm{a}}$ calculated utilizing the correct *a priori* covariance and measurement noise covariances. The key to our development is the observation that unbiased linear estimates of X can be written in an incremental correction format

$$\begin{bmatrix} \hat{X} \\ \hat{y} \end{bmatrix} = \begin{bmatrix} \tilde{X} \\ \tilde{y} \end{bmatrix} + \begin{bmatrix} K_X \\ 0 \end{bmatrix} \left(z - [A_X \; A_y] \begin{bmatrix} \tilde{X} \\ \tilde{y} \end{bmatrix} \right) \tag{2.4}$$

where K_X is an arbitrary gain matrix and

$$z = A_X X + A_y y + \nu, \qquad E(\nu) = 0, \qquad E(\nu\nu^{\mathrm{T}}) = P_\nu^{\mathrm{a}} \tag{2.5}$$

We assume that the measurement is normalized, having been premultiplied by R_ν (where $P_\nu = R_\nu^{-1} R_\nu^{-T}$ is the assumed *a priori* measurement covariance). Thus the ν appearing in (2.5) has assumed unit covariance and an actual covariance $P_\nu^{\mathrm{a}} = R_\nu \bar{P}_\nu^{\mathrm{a}} R_\nu^{\mathrm{T}}$, where $\bar{P}_\nu^{\mathrm{a}}$ is the measurement covariance before R_ν scaling.

Representation (2.4) is, as we proceed to show, a direct result of requiring the estimate to be linear and unbiased with y stipulated as a vector of consider variables. First, linearity of the estimate requires

$$\begin{bmatrix} \hat{X} \\ \hat{y} \end{bmatrix} = L_1 \begin{bmatrix} \tilde{X} \\ \tilde{y} \end{bmatrix} + L_2 z \tag{2.6}$$

and unbiasedness of the estimate implies

$$E\begin{bmatrix} \hat{X} \\ \hat{y} \end{bmatrix} = E\begin{bmatrix} \tilde{X} \\ \tilde{y} \end{bmatrix} = \begin{bmatrix} X \\ y \end{bmatrix}$$

These two relations give

$$\begin{bmatrix} X \\ y \end{bmatrix} = L_1 \begin{bmatrix} X \\ y \end{bmatrix} + L_2 [A_X \; A_y] \begin{bmatrix} X \\ y \end{bmatrix}$$

and since $\begin{bmatrix} x \\ y \end{bmatrix}$ is arbitrary this requires that

$$L_1 = I - L_2[A_X \; A_y]$$

Using this relationship to eliminate L_1 from (2.6) gives the desired result with

$$L_2 = \begin{bmatrix} K_X \\ 0 \end{bmatrix} \begin{matrix} \}N_X \\ \}N_y \end{matrix}$$

The reason for the lower zero block is that the consider parameters y are not to be estimated, i.e., we require $\hat{y} = \tilde{y}$.

The gain K_X is as yet unspecified. Appropriate choices of K_X will yield various weighted least squares filters commonly used. Other choices of K_X may also yield useful results.

To arrive at the covariance of the estimate $\hat{X}$, we rewrite (2.4) in the form of an "error" equation and utilize the measurement equation (2.5):

$$\begin{bmatrix} X - \hat{X} \\ y - \hat{y} \end{bmatrix} = \left(I - \begin{bmatrix} K_X \\ 0 \end{bmatrix} [A_X \; A_y] \right) \begin{bmatrix} X - \tilde{X} \\ y - \tilde{y} \end{bmatrix} + \begin{bmatrix} K_X \\ 0 \end{bmatrix} \nu \tag{2.7}$$

Remark: The term "error equation" is used because the terms $X - \overline{X}$, $y - \overline{y}$, and ν are errors.

"Squaring" Eq. (2.7), by multiplying by its matrix transpose, and taking expectations leads to the basic covariance results

$$\hat{P}_X^{\mathrm{a}} = (I - K_X A_X)\tilde{P}_X^{\mathrm{a}}(I - K_X A_X)^{\mathrm{T}} + K_X\left(P_\nu^{\mathrm{a}} + A_y P_y^{\mathrm{a}} A_y^{\mathrm{T}}\right)K_X^{\mathrm{T}}$$
$$- \left[(I - K_X A_X)\tilde{P}_{Xy}^{\mathrm{a}} A_y^{\mathrm{T}} K_X^{\mathrm{T}}\right] - \left[(I - K_X A_X)\tilde{P}_{Xy}^{\mathrm{a}} A_y^{\mathrm{T}} K_X^{\mathrm{T}}\right]^{\mathrm{T}} \tag{2.8}$$

and

$$\hat{P}_{Xy}^{\mathrm{a}} = (I - K_X A_X)\tilde{P}_{Xy}^{\mathrm{a}} - K_X A_y P_y^{\mathrm{a}} \tag{2.9}$$

Because y is not estimated,

$$\hat{P}_y^{\mathrm{a}} = \tilde{P}_y^{\mathrm{a}} = P_y^{\mathrm{a}}$$

Equation (2.8) is the *general batch filter consider covariance*, and Eq. (2.9) shows how the correlation of the estimated and consider parameters is affected by processing the data z.

Special Consider Filters of General Interest

A. *Filter Uses Only Current Data; No a Priori or Consider Variables* This filter is the familiar least squares filter (cf. Eq. (II.2.4))

$$\hat{X}_{ls} = (A_X^T A_X)^{-1} A_X^T z = Fz \tag{2.10}$$

Note that the reason (2.10) is consistent with (2.4) is that K_X is the coefficient of z, so that $K_X = F$ and the coefficient of the *a priori* estimate is zero, i.e., $K_X A_X = I$. This is as it should be since the covariance of the *a priori* estimate is infinite. The computed and true covariances of $\hat{X}_{ls}$ are (cf. Eq. (2.8) with $I - K_X A_X = 0$ and $P_y^a = 0$)

$$\hat{P}_{ls} = (A_X^T A_X)^{-1} \tag{2.11}$$

and

$$\hat{P}_{ls}^a = F P_\nu^a F^T = \hat{P}_{ls}(A_X^T P_\nu^a A_X)\hat{P}_{ls} \tag{2.12}$$

The consider covariance $\hat{P}_{ls}^a$ is the true covariance corresponding to a batch filter, with no *a priori* estimate, that uses wrong data weighting.†

B. *Filter Uses Current Data and an a Priori Estimate; No Consider Variables* In this case K_X has the form of a Kalman gain and can be written as

$$K_X = \tilde{P}_X A_X^T \left[A_X \tilde{P}_X A_X^T + I \right]^{-1} = \left[\tilde{P}_X^{-1} + A_X^T A_X \right]^{-1} A_X^T \qquad \text{(cf. p. 21)} \tag{2.13}$$

Since $P_y^a = 0$ the fundamental result (2.8) reduces to

$$\hat{P}_X^a = (I - K_X A_X)\tilde{P}_X^a (I - K_X A_X)^T + K_X P_\nu^a K_X^T \tag{2.14}$$

In some applications it is convenient to have this result, (2.13) and (2.14), expressed in a form analogous to (2.12). To accomplish this we note that a modest manipulation of Eq. (2.13) gives

$$I - K_X A_X = \left(\tilde{P}_X^{-1} + A_X^T A_X \right)^{-1} \tilde{P}_X^{-1}$$

Substituting this expression and (2.13) into (2.14) results in

$$\hat{P}_{kf}^a = \hat{P}_X^a = \hat{P}_X \left[\left(\tilde{P}_X \right)^{-1} \tilde{P}_X^a \left(\tilde{P}_X \right)^{-1} + A_X^T P_\nu^a A_X \right] \hat{P}_X \tag{2.15}$$

†It is prudent to remind the reader that our filter formulation assumes that the measurement covariance is unity.

where

$$\hat{P}_{\mathrm{kf}} = \hat{P}_X = (I - K_X A_X)\tilde{P}_X = \left[\tilde{P}_X^{-1} + A_X^{\mathrm{T}} A_X\right]^{-1} \tag{2.16}$$

The subscript "kf" is used because K_X is the Kalman filter gain and the first of Eqs. (2.16) is the measurement update equation associated with Kalman filtering.

The result (2.15) highlights the effect of errors in the *a priori* state and measurement covariances, $\tilde{P}_X^{\mathrm{a}}$ and P_ν^{a}, respectively. Note how (2.15) generalizes the least squares result of case A.

C. *The Standard Consider Covariance* The standard consider covariance consists of a filter that uses correct *a priori* for the estimated parameters, but neglects to include the y consider parameters. Evidently K_X and $\hat{P}_X$ are the same as (2.13) and (2.16), respectively, with $\tilde{P}_X = \tilde{P}_X^{\mathrm{a}}$. Inserting these terms into (2.8), assuming $\tilde{P}_{Xy}^{\mathrm{a}} = 0$, and turning the crank gives

$$\hat{P}_{\mathrm{CON}}^{\mathrm{a}} = \hat{P}_X + \left(\hat{P}_X A_X^{\mathrm{T}} A_y\right) P_y^{\mathrm{a}} \left(\hat{P}_X A_X^{\mathrm{T}} A_y\right)^{\mathrm{T}} \tag{2.17}$$

This result corresponds with Eq. (VII.2.18). Comparing these expressions, we conclude that the sensitivity $\hat{S}$ is given by

$$\hat{S} = -\hat{R}_X^{-1}\hat{R}_{Xy} = -\hat{P}_X A_X^{\mathrm{T}} A_y \tag{2.18}$$

This result appears again in Chapter IX.

D. *Curkendall's Generalized Consider Covariance* Curkendall [2] suggested considering Problems B and C simultaneously. This problem is more realistic than either problem separately because one is tempted to compensate for the neglected parameters by modifying the *a priori* on the estimated parameters. The resulting generalized consider covariance is

$$P_{\mathrm{G.CON}}^{\mathrm{a}} = \hat{P}_{kf}^{\mathrm{a}} + \left(\hat{P}_X A_X^{\mathrm{T}} A_y\right) P_y^{\mathrm{a}} \left(\hat{P}_X A_X^{\mathrm{T}} A_y\right)^{\mathrm{T}} \tag{2.19}$$

Here too it is assumed that X and y are independent *a priori*.

E. *Schmidt Type Consider Filter* The Schmidt type filter neglects the correlations between the estimated and consider parameters, but assumes that the filter uses correct *a priori* statistics for the estimated parameter covariance, consider parameter covariance, and measurement covariance. The gain K_X is chosen to minimize the *a posteriori* covariance trace (minimum variance); see Chapter II. The result, obtainable by setting the gradient of Eq. (2.8) to zero, is

$$K_{\mathrm{S.CON}} = \left(\tilde{P}_X^{-1} + A_X^{\mathrm{T}} P_{\mathrm{EQUIV}}^{-1} A_X\right)^{-1} A_X^{\mathrm{T}} P_{\mathrm{EQUIV}}^{-1} \tag{2.20}$$

where S.CON refers to Schmidt consider and

$$P_{\text{EQUIV}} = I + A_y P_y^{\text{a}} A_y^{\text{T}} \tag{2.21}$$

i.e., $\nu + A_y y$ is considered as the noise term. Thus this problem with consider variables is equivalent to a problem using the filter of type B with a newly defined noise covariance P_{EQUIV} instead of P_ν. Perusal of Eq. (2.8) with $\tilde{P}_{Xy}^{\text{a}} = 0$ makes it apparent that P_{EQUIV} plays the role of a generalized noise covariance.

A generalized form of the Schmidt consider filter retains the correlations. It seems as if this kind of filter, if applied sequentially, would be in some sense a minimum variance estimator of X. Indeed, Jazwinski [3] asserts that this is, in fact, true. To see that the assertion of optimality of this filter is not true, except perhaps when using a nonsensically restrictive interpretation, consider the following situation. Suppose two batches of observations are sequentially processed using the generalized Schmidt consider filter. When the first batch is processed, X will be optimally estimated; but y is deliberately set to zero and thereby information is destroyed. Processing of the second batch of data can use the X–y correlations induced in the previous step, but there is no y estimate based on the first data batch for use in estimating X. On the other hand, suppose that both batches of data were processed simultaneously with X estimated and y considered. In this case X will be optimally estimated (minimum variance), and the result is better (in the minimum variance sense) than the two-step consider filter result. Observe that neither filter estimates y, i.e., both are consider filters.

Note that "nonoptimality" of the Schmidt filter is not an indictment of the method. The Schmidt consider filter has in various instances successfully avoided the filter divergence problem mentioned in the introduction.

VIII.3 Considering the Effects of Unestimated Parameters in the Kalman Filter

In this section we summarize the covariance evaluation formulae that are used for Kalman filter performance analyses. Our motive for including the Kalman error analysis exposition is in large part to compare the Kalman and SRIF evaluation algorithms. A second reason is our desire to point up an often overlooked feature of the Kalman evaluation which allows one to compute sensitivities corresponding to unestimated bias parameters.

The performance analysis to be discussed will involve evaluating the effects of

(1) an incorrect *a priori* covariance on the estimated state vector,
(2) an incorrect *a priori* covariance on the measurement vector,
(3) an incorrect *a priori* covariance on the colored noise parameters and on the process noise, and
(4) unestimated biases y_c and colored noise p_c, i.e., considering bias and colored noise parameters.

Our analysis here, for the first two items, extends the work introduced in Section VIII.2, because mapping effects were not included there.

The model used to construct our filter estimate is the one described in Chapter VII, Eqs. (VII.1.1) and (VII.1.2), and we let

$$X = \begin{cases} p & \text{correlated process noise, } N_p \text{ in number} \\ x & \text{states that are time varying, but do not explicitly depend on white process noise, } N_x \text{ in number} \\ y & \text{bias parameters, } N_y \text{ in number} \end{cases}$$

For brevity we use X wherever possible, and will omit explicit reference to p, x, and y in this chapter. For example, we say that X is to be estimated using the filter algorithm of Chapter VII.

Remark: The X model is adequate for a rather wide range of applications, especially since the assumption of a diagonal process noise dynamics matrix M in Eq. (VII.1.2) is not an essential requirement. Note that the presence of process noise terms in the x model equation (cf. Eq. (VII.1.1)) is not ruled out because one can use p's that are white noise, with corresponding m's that are zero.

In addition to incorrect *a priori* covariances associated with X (X itself, the white process noise w, and the measurement noise ν), we include vectors y_c of bias parameters and p_c of colored noise which are to be *considered* in the analysis—i.e., these parameters are assumed to be present, but are not included in the filter model. Notation and relations associated with y_c and p_c follow:

$$y_c = \begin{bmatrix} y_c(1) \\ \vdots \\ y_c(N_{y_c}) \end{bmatrix}; \qquad p_c = \begin{bmatrix} p_c(1) \\ \vdots \\ p_c(N_{p_c}) \end{bmatrix} \tag{3.1}$$

i.e., y_c and p_c without an argument are vectors, while an integer argument represents a vector component. Both vectors are assumed to be zero mean, and the covariance of p_c is in steady state (cf. Appendix VII.A). Associated with $p_c(j)$ is an exponential decay factor.

$$m_c(j) = \exp(-\Delta t/\tau_c(j))$$

and a zero mean white noise process $w_c(j)$ with variance

$$\sigma_{w_c}^2(j) = E(w_c^2(j)) = 1/r_{w_c}^2(j) = \sigma_c^2(j)(1 - m_c^2(j)) \tag{3.2}$$

where $\sigma_c^2(j)$ is the *a priori* variance of $p_c(j)$.

Because of the steady-state assumption the time dependence of p_c is omitted, and this simplifies notation. When, however the propagation from t_j to t_{j+1} is discussed there will be an ambiguity about which time p_c is referred to. Note that although the statistics of p_c are constant, p_c is time varying. To avoid confusion in this instance we write $p_{c\,|\,t_j}$; and if the kth component is referred to, we write $p_{c\,|\,t_j}(k)$.

The state vector for our filter error analysis is arranged in the form of

$$\begin{bmatrix} p_c \\ X \\ y_c \end{bmatrix} \tag{3.3}$$

This arrangement is premeditated. For our covariance analyses it is convenient to have the y_c terms as an addendum to the filter. We also find it expedient to have all the bias parameters in tandem. It turns out that the SRIF analyses are also benefited from this arrangement of parameters, but that will be discussed in the next chapter.

The measurement and dynamic models are written as

$$z_j = A_{p_c} p_{c\,|\,t_j} + A_X X_j + A_{y_c} y_c + \nu_j \tag{3.4}$$

$$\begin{bmatrix} p_c \\ X \\ y_c \end{bmatrix}_{j+1} = \begin{bmatrix} M_c & 0 & 0 \\ \Phi_{Xp_c} & \Phi_X & \Phi_{Xy_c} \\ 0 & 0 & I \end{bmatrix} \begin{bmatrix} p_c \\ X \\ y_c \end{bmatrix}_j + \begin{bmatrix} w_c \\ G_X w \\ 0 \end{bmatrix}_j \tag{3.5}$$

Statistics relevant to the covariance analysis follow;

P_X Filter covariance of the estimated parameters.

$\begin{bmatrix} P_{p_c} & P_{p_cX}^a \\ P_{Xp_c}^a & P_X^a \end{bmatrix}$ Actual covariance of the $\begin{bmatrix} p_c \\ X \end{bmatrix}$ model. (Note that $P_{p_c} = P_{p_c}^a$.)

Remark: Matrices with a superscript "a" refer to *actual* values, i.e., values believed to be, in some sense, correct. Quantities without this designation are filter values that are assumed, formal statistics and are often chosen via ad hoc reasoning. The filter (formal) and actual statistics often differ markedly.

P_{y_c} Actual covariance for the y_c bias parameters.

P_{y_c} is generally diagonal and initially y_c is uncorrelated with the filter estimates of p_c and X. These restrictions are not essential but we believe that it is preferable not to burden the development with excess generality.

Remark: The filter assumes $P_{p_c} = 0$ and $P_{y_c} = 0$ and these variables are omitted from the filter model.

P_w, P_w^a Filter and actual *a priori* covariances for the process noise driving the stochastic parameters of X.

P_v, P_v^a Filter and actual values for the measurement noise. In our analyses, P_v is generally assumed to be the identity.

Corresponding to measurement process (3.4) and model (3.5), we have the Kalman estimator

$$\hat{X}_j = \tilde{X}_j + K_j(z_j - A_X\tilde{X}_j) \tag{3.6}$$

$$\tilde{X}_{j+1} = \Phi_X\hat{X}_j \tag{3.7}$$

where K_j is the filter gain. Various methods for selecting gains were suggested in Section VIII.2. In general these gains are generated by a filter covariance recursion (cf. Section VI.3). We do not concern ourselves here with this covariance, except to note that the use of incorrect *a priori* or the presence of unestimated parameters is reflected in the gain sequence $\{K_j\}$.

We find it convenient to decompose the model as follows:

$$X_j = X_j^p + X_j^{yc} \tag{3.8}$$

where X^p and X^{yc} are defined by

$$\begin{bmatrix} p_c \\ X^p \end{bmatrix}_{j+1} = \begin{bmatrix} M_c & 0 \\ \Phi_{Xp_c} & \Phi_X \end{bmatrix} \begin{bmatrix} p_c \\ X^p \end{bmatrix}_j + \begin{bmatrix} w_{c\,|\,t_j} \\ G_X w_j \end{bmatrix} \tag{3.9}$$

and

$$X_{j+1}^{yc} = \Phi_X X_j^{yc} + \Phi_{Xy_c} y_c \tag{3.10}$$

where

$$X_0^p = X_0, \qquad X_0^{yc} = 0, \qquad P_{X\,|\,t_0}^p = P_X^a, \qquad P_{X\,|\,t_0}^{yc} = 0$$

By introducing this decomposition we separate the error contributions; X^p involves all X-related error sources except those due to the unmodeled y_c parameters and X^{yc} involves only the error contributions due to y_c. The decomposition is especially useful when there are a large number of y_c consider parameters. The covariance decomposition we have in mind comes about by separating the estimate errors in a fashion analogous to (3.9) and (3.10) and then forming the appropriate covariance matrices.

Measurement Processing (at time t_j)

Combining the filter update (3.6) with the measurement (3.4) and arranging the results in a form compatible with model (3.9) gives the following estimate error representation:

$$\begin{bmatrix} p_c - \hat{p}_c \\ X - \hat{X} \end{bmatrix}_j = \begin{bmatrix} I & 0 \\ -K_j A_{p_c} & I - K_j A_X \end{bmatrix} \begin{bmatrix} p_c - \tilde{p}_c \\ X - \tilde{X} \end{bmatrix}_j - \begin{bmatrix} 0 \\ K_j \nu_j \end{bmatrix} - \begin{bmatrix} 0 \\ K_j A_{y_c} y_c \end{bmatrix}$$

Substituting the decomposition (3.8) into the bottom row of this equation gives

$$X_j^{yc} + \left(X_j^p - \hat{X}_j\right) = (I - K_j A_X) X_j^{yc} - K_j A_{y_c} y_c + (I - K_j A_X)\left(X_j^p - \tilde{X}_j\right) \\ - K_j A_{p_c} (p_c - \tilde{p}_c)_{t_j} - K_j \nu_j \tag{3.11}$$

To extract an equation for the error $X^p - \overline{X}$† that is independent of y_c we introduce an auxiliary vector $\tilde{X}^{yc}$

$$\tilde{X} = \tilde{X}^p + \tilde{X}^{yc} \qquad \text{(cf. Eq. (3.8))}$$

and decompose (3.11) into two equations:

$$\begin{bmatrix} \Delta_{\hat{p}_c} \\ \Delta \hat{X}^p \end{bmatrix} = \begin{bmatrix} I & 0 \\ -K_j A_{p_c} & I - K_j A_X \end{bmatrix} \begin{bmatrix} \Delta \tilde{p}_c \\ \Delta \tilde{X}^p \end{bmatrix}_j - \begin{bmatrix} 0 \\ K_j \nu_j \end{bmatrix} \tag{3.12}$$

$$\Delta \hat{X}_j^{yc} = (I - K_j A_X) \Delta \tilde{X}_j^{yc} - K_j A_{y_c} y_c \tag{3.13}$$

where

$$\Delta \bar{p}_c = p_c - \bar{p}_c, \qquad \Delta \overline{X}^p = X^p - \overline{X}^p, \text{ etc.}$$

†The overbar means either ^ or ~.

Propagation (t_j to t_{j+1})

Time propagation corresponding to the measurement updates (3.12) and (3.13) is a direct consequence of the filter propagation (3.9) and (3.10) and the model (3.7), and we write

$$\begin{bmatrix} \tilde{p}_c \\ \tilde{X}^p \end{bmatrix}_{j+1} = \begin{bmatrix} M_c & 0 \\ \Phi_{Xp_c} & \Phi_X \end{bmatrix} \begin{bmatrix} \hat{p}_c \\ \hat{X}^p \end{bmatrix}_j, \quad \tilde{X}^{yc}_{j+1} = \Phi_X \hat{X}^{yc}_j \tag{3.14}$$

Thus, differencing the model and the estimate we obtain

$$\begin{bmatrix} \Delta\tilde{p}_c \\ \Delta\tilde{X}^p \end{bmatrix}_{j+1} = \begin{bmatrix} M_c & 0 \\ \Phi_{Xp_c} & \Phi_X \end{bmatrix} \begin{bmatrix} \Delta_{p_c} \\ \Delta\tilde{X}^p \end{bmatrix}_j + \begin{bmatrix} w_{c\,|\,t} \\ G_X w \end{bmatrix}_j \tag{3.15}$$

$$\Delta\tilde{X}^{yc}_{j+1} = \Phi_X\, \Delta\hat{X}^{yc}_j + \Phi_{Xy_c} y_c \tag{3.16}$$

The covariance recursions for $\bar{P}^p_X = E[(\Delta\bar{X}^p)(\Delta\bar{X}^p)^{\mathrm{T}}]$ are direct consequences of Eqs. (3.12) and (3.16):

$$\begin{bmatrix} P_{p_c} & \hat{P}^p_{p_cX} \\ \hat{P}^p_{Xp_c} & \hat{P}^p_X \end{bmatrix}_j = \begin{bmatrix} I & 0 \\ -K_jA_{p_c} & I - K_jA_X \end{bmatrix} \begin{bmatrix} P_{p_c} & \tilde{P}^p_{p_cX} \\ \tilde{P}^p_{Xp_c} & \tilde{P}^p_X \end{bmatrix}_j$$

$$\times \begin{bmatrix} I & 0 \\ -K_jA_{p_c} & I - K_jA_X \end{bmatrix}^{\mathrm{T}} + \begin{bmatrix} 0 & 0 \\ 0 & K_jP^{\mathrm{a}}_{\nu}K_j^{\mathrm{T}} \end{bmatrix} \tag{3.17}^{\dagger}$$

i.e.,

$$\hat{P}^p_X(j) = (I - K_jA_X)\tilde{P}^p_X(j)(I - K_jA_X)^{\mathrm{T}} + K_j\left(A_{p_c}P_cA^{\mathrm{T}}_{p_c} + P^{\mathrm{a}}_{\nu}\right)K_j^{\mathrm{T}}$$

$$-\left[(I - K_jA_X)\tilde{P}^p_{Xp_c}(j)A^{\mathrm{T}}_{p_c}K_j^{\mathrm{T}}\right] - \left[(I - K_jA_X)\tilde{P}^p_{Xp_c}(j)A^{\mathrm{T}}_{p_c}K_j^{\mathrm{T}}\right]^{\mathrm{T}} \tag{3.18}$$

$$\hat{P}^p_{Xp_c}(j) = (I - K_jA_X)\tilde{P}^p_{Xp_c}(j) - K_jA_{p_c}P_c \tag{3.19}$$

† $P_c = \hat{P}_{c\,|\,t_j} = \tilde{P}_{c\,|\,t_j}$ since the colored noise is in steady state.

Remark: The covariance relations (3.18) and (3.19) should be evaluated using the computationally efficient vector outer product (cf. Section V.2).

$$\begin{bmatrix} P_{p_c} & \tilde{P}^p_{p_c X} \\ \tilde{P}^p_{X p_c} & \tilde{P}^p_X \end{bmatrix}_{j+1} = \begin{bmatrix} M_c & 0 \\ \Phi_{X p_c} & \Phi_X \end{bmatrix} \begin{bmatrix} P_{p_c} & \hat{P}^p_{p_c X} \\ \hat{P}_{X p_c} & \hat{P}^p_X \end{bmatrix}_j \begin{bmatrix} M_c & 0 \\ \phi_{X p_c} & \Phi_X \end{bmatrix}^{\mathrm{T}} + \begin{bmatrix} P_{w_c} & 0 \\ 0 & G_X P^{\mathrm{a}}_w G_X^{\mathrm{T}} \end{bmatrix} \tag{3.20}$$

i.e.,

$$\tilde{P}^p_X(j+1) = \Phi_X \hat{P}^p_X(j) \Phi_X^{\mathrm{T}} + \Phi_{X p_c} P_c \Phi_{X p_c}^{\mathrm{T}} + G_X P^{\mathrm{a}}_w G_X^{\mathrm{T}} + \left[\Phi_X \hat{P}^p_{X p_c}(j) \Phi_{X p_c}^{\mathrm{T}}\right] + \left[\Phi_X \hat{P}^p_{X p_c}(j) \Phi_{X p_c}^{\mathrm{T}}\right]^{\mathrm{T}} \tag{3.21}$$

$$\tilde{P}^p_{X p_c}(j+1) = \left(\Phi_X \hat{P}^p_{X p_c}(j) + \Phi_{X p_c} P_c\right) M_c^{\mathrm{T}} \tag{3.22}$$

Equations (3.18)–(3.19) and (3.21)–(3.22) are the measurement and time update recursions for P^p_X.

Remark: One could apply U–D factorization to the covariance recursions (3.17) and (3.20). The resulting algorithms, discussed in Thornton (1976), are accurate and computationally efficient.

Perusal of Eqs. (3.13) and (3.16) shows that X^{y_c} is a linear function of y_c, and we write

$$\bar{X}_j^{y_c} = \overline{\mathrm{Sen}}_X(j) y_c \tag{3.23}$$

Substituting (3.23) back into the X^{y_c} recursions results in the $\overline{\mathrm{Sen}}$ matrix measurement and time update recursions

$$\widehat{\mathrm{Sen}}_X(j) = (I - K_j A_X)\, \widetilde{\mathrm{Sen}}_X(j) - K_j A_{y_c} \tag{3.24}$$

$$\widetilde{\mathrm{Sen}}_X(j+1) = \Phi_X\, \widehat{\mathrm{Sen}}_X(j) + \Phi_{X y_c} \tag{3.25}$$

The Sen_X result indicates how to consider the effects of unestimated biases. The columns of Sen_X represent 1 sigma perturbations due to the presence of unestimated and uncorrelated y_c bias parameters. Observe that the Sen_X computation is an almost trivial addendum to the filter algorithm. Comparing the covariance and SRIF sensitivity calculations (cf. Section VII.2), we conclude that both are addenda to the filter computation. The SRIF, however, has the feature of being able to decide *a posteriori* to estimate the biases. In the case of the Kalman process, the decision to estimate the biases necessitates re-solving the entire problem.†

To obtain the actual error covariance $\bar{P}_X = E[(\Delta\bar{X})(\Delta\bar{X})^T]$, observe that the decomposition (3.8) has resulted in

$$X - \bar{X} = (X^p - \bar{X}^p) + (X^{y_c} - \bar{X}^{y_c})$$

and that by construction the terms are statistically independent. Thus, the actual error covariance is given by‡

$$\bar{P}_X = \operatorname{cov}(\bar{X}) = \bar{P}_X^p + \overline{\text{Sen}}_X P_{y_c} \overline{\text{Sen}}_X^{\text{T}} \tag{3.26}$$

One could use the relations defined here for a SRIF evaluation. All that need be done is to construct the gain, K, referring to Section VIII.2. We do not in general recommend such a procedure, because we believe that the augmented information array discussed in the next chapter is generally a more efficient method. When, however, actual covariances are needed with great frequency one might consider using the covariance method just described.

VIII.4 State Variable Augmentation to Evaluate the Effects of Mismodeled Colored Noise

We show in this section that the software developed to analyze the effects of incorrect *a priori* and unmodeled parameters can be used, without modification, to analyze the effects of mismodeled colored noise. Our approach consists of adding a state variable for each mismodeled colored noise parameter. To keep the notation from becoming too cluttered, we omit consider parameters from this development. That our result is valid when consider parameters are present can be demonstrated by interpreting the state vector X so as to include the consider parameters. The problem can be succinctly stated as follows.

Suppose that we are given the linear system

$$\begin{bmatrix} X \\ p \end{bmatrix}_{j+1} = \begin{bmatrix} \Phi_X & \Phi_{Xp} \\ 0 & M \end{bmatrix} \begin{bmatrix} X \\ p \end{bmatrix}_j + \begin{bmatrix} G_X w \\ w_p \end{bmatrix}_j \tag{4.1}$$

and the observations

$$z_j = A_X X_j + A_p p_j + \nu_j \tag{4.2}$$

†One could salvage the *a posteriori* bias estimation capability if the covariance algorithm were formulated using the U–D factorization (cf. Chapters V–VII).

‡If X is a random vector and $\bar{X}$ is an estimate of it, then $\operatorname{cov}(\bar{X}) \triangleq E[(X - \bar{X})(X - \bar{X})^{\text{T}}]$.

An estimator for this system is given by

$$\begin{bmatrix} \hat{X} \\ \hat{p} \end{bmatrix}_j = \begin{bmatrix} \tilde{X} \\ \tilde{p} \end{bmatrix}_j + \begin{bmatrix} K_X \\ K_p \end{bmatrix}_j \left(z_j - A_X\tilde{X}_j - A_p\tilde{p}_j\right) \tag{4.3}$$

$$\begin{bmatrix} \tilde{X} \\ \tilde{p} \end{bmatrix}_{j+1} = \begin{bmatrix} \Phi_X & \Phi_{Xp} \\ 0 & M' \end{bmatrix} \begin{bmatrix} \hat{X} \\ \hat{p} \end{bmatrix}_j \tag{4.4}$$

The gains K_X and K_p depend on assumed (filter) statistics for $\tilde{X}_0$, $\tilde{p}_0$, $\{w_j\}$, $\{w_p(j)\}$, $\{v_j\}$ and the assumed filter model dynamics, Φ_X, Φ_{Xy}, and M'. The difference between this problem and the one described in the previous section is that in the present problem the estimate prediction equation (4.4) differs from the model (4.1).

We will show that the covariance of the X estimate can be computed from an augmented state filter using the following inputs:

$$P_0^{\mathrm{a}} = \begin{bmatrix} \mathrm{cov}(\tilde{X}_0) & 0 & E\left[(\Delta\tilde{X}_0)(\Delta\tilde{p}_0)^{\mathrm{T}}\right] \\ & 0 & 0 \\ & & \mathrm{cov}(\tilde{p}_0) \end{bmatrix} \tag{4.5a}$$

$$Q_{\mathrm{TOT}}^{\mathrm{a}} = \mathrm{Diag}\left(E(ww^{\mathrm{T}})\ 0,\ E\left(w_p w_p^{\mathrm{T}}\right)\right), \qquad P_v^{\mathrm{a}} = E(vv^{\mathrm{T}}) \tag{4.5b}$$

$$K_{\mathrm{TOT}} = \begin{bmatrix} K_X \\ K_p \\ 0 \end{bmatrix}, \qquad \Phi_{\mathrm{TOT}} = \begin{bmatrix} \Phi_X & \Phi_{Xp} & \Phi_{Xp} \\ 0 & M' & 0 \\ 0 & 0 & M \end{bmatrix} \tag{4.5c}$$

The upper left block of the covariance P^{a} is the covariance of the X estimate.

For analysis purposes one may view the filter as operating on the same system, (4.5), but with different *a priori* statistics P_0, Q_{TOT}, and P_v, where

$$P_0 = \begin{bmatrix} P_X & P_{Xp} & 0 \\ & P_p & 0 \\ & & 0 \end{bmatrix}, \qquad Q_{\mathrm{TOT}} = \mathrm{Diag}(Q_w, Q_p, 0) \tag{4.6}$$

The P's and Q's are assumed filter statistics.

The importance of Eqs. (4.5) and (4.6) is that they demonstrate that the problem of evaluating the effects of mismodeled colored noise dynamics can be viewed as a problem with incorrect *a priori* statistics. Consequently the results of Section VIII.3 are applicable. To understand how our assertion (4.5) comes about one has only to consider the system (4.7) and (4.8):

$$\begin{bmatrix} X \\ p' \\ p \end{bmatrix}_{j+1} = \begin{bmatrix} \Phi_X & \Phi_{Xp} & \Phi_{Xp} \\ 0 & M' & 0 \\ 0 & 0 & M \end{bmatrix} \begin{bmatrix} X \\ p' \\ p \end{bmatrix}_j + \begin{bmatrix} G_X w \\ 0 \\ w_p \end{bmatrix}_j \tag{4.7}$$

with $p'(0) = 0$

$$z_j = A_X X_j + A_p p_j' + A_p p_j + v_j \tag{4.8}$$

Remark: Observe that p_j' equals zero and is an artifice to facilitate our manipulations.

To develop recursive equations for the error in the X estimate we introduce estimate equations compatible with (4.7) and (4.8):

$$\begin{bmatrix} \hat{X} \\ \hat{p}' \\ \hat{p} \end{bmatrix}_j = \begin{bmatrix} \tilde{X} \\ \tilde{p}' \\ \tilde{p} \end{bmatrix}_j + \begin{bmatrix} K_X \\ K_p \\ 0 \end{bmatrix}_j \left[z_j - \left(A_X \tilde{X}_j + A_p \tilde{p}_j' + A_p \tilde{p}_j \right) \right] \tag{4.9}$$

$$\begin{bmatrix} \tilde{X} \\ \tilde{p}' \\ \tilde{p} \end{bmatrix}_{j+1} = \begin{bmatrix} \Phi_X & \Phi_{Xp} & \Phi_{Xp} \\ 0 & M' & 0 \\ 0 & 0 & M \end{bmatrix} \begin{bmatrix} \hat{X} \\ \hat{p}' \\ \hat{p} \end{bmatrix}_j \tag{4.10}$$

with $\tilde{p}_0 = 0$.

Remark: Observe that $\tilde{p}_j$ and $\hat{p}_j$ are zero and they too are an artifice to facilitate our manipulations.

Forming error equations in the same fashion as described in Sections VIII.2 and VIII.3 gives the desired results, which we express as

$$\hat{\Delta}_j = (I - K_{\mathrm{TOT}} A_{\mathrm{TOT}})\, \tilde{\Delta}_j - K_{\mathrm{TOT}} v_j \tag{4.11}$$

$$\tilde{\Delta}_{j+1} = \Phi_{\mathrm{TOT}}\, \hat{\Delta} + w_{\mathrm{TOT}}(j) \tag{4.12}$$

where

$$\bar{\Delta}_j = \begin{bmatrix} X - \bar{X} \\ p' - \bar{p}' \\ p - \bar{p} \end{bmatrix}_j, \qquad \text{"–" equals "~" or "^"}$$

$$A_{\mathrm{TOT}} = [A_X \; A_p \; A_p], \qquad w_{\mathrm{TOT}}(j) = \begin{bmatrix} G_X w \\ 0 \\ w_p \end{bmatrix}_j$$

The covariance recursions can be obtained by "squaring" the error recursions (4.11) and (4.12) and applying the expectation operator. Note that the upper left covariance elements correspond to $E[(X-\bar{X})(X-\bar{X})^{\mathrm{T}}]$. The other elements in the matrix are also useful. For example, the covariance of the estimate for p, which we have called $\bar{p}'$, is obtainable by noting that

$$\bar{\Delta}_j = \begin{bmatrix} X - \bar{X} \\ -\bar{p}' \\ p \end{bmatrix}_j = \begin{bmatrix} \bar{\Delta}(1) \\ \bar{\Delta}(2) \\ \bar{\Delta}(3) \end{bmatrix}_j$$

and

$$p_j - \bar{p}_j' = \bar{\Delta}_j(3) + \bar{\Delta}_j(2) \tag{4.13}$$

Thus

$$\operatorname{cov}(\bar{p}_j') \equiv E\left[(p_j - \bar{p}_j')(p_j - \bar{p}_j')^{\mathrm{T}}\right] = E[\Delta_j(2)\,\Delta_j^{\mathrm{T}}(2)] + E\left[\bar{\Delta}_j(3)\,\bar{\Delta}_j^{\mathrm{T}}(3)\right]$$
$$+ E\left[\bar{\Delta}_j(2)\,\bar{\Delta}_j^{\mathrm{T}}(3)\right] + E\left[\bar{\Delta}_j(2)\,\bar{\Delta}_j^{\mathrm{T}}(3)\right]^{\mathrm{T}}$$

and the expectations involving $\bar{\Delta}$ are elements of $\bar{P}^{\mathrm{a}}$.

Because of (4.13) it happens that when $\bar{p}_j'$ is a good estimate of p_j then $\bar{P}^{\mathrm{a}}$ will be near singular. Our experiences are limited, but no computational problems have been observed.

Acknowledgment: The idea of treating colored noise mismodeling as a problem with incorrect *a priori* was first suggested to the author by Mr. Ken Rourke of the Jet Propulsion Laboratory (Bierman and Rourke (1972)).

References

[1] Sage, A. P., and Melsa, J. L., "Estimation Theory with Applications to Communications and Control." McGraw-Hill, New York, 1971.

The authors present lengthy but very general error analysis algorithms. The derivations are easy to follow, but because of their generality the algorithms are unwieldy in some applications. Our sensitivity algorithms, which differ from those presented in this text, are better suited for computation.

[2] Curkendall, D. W., Toward the design of more rational filters—a generalized consider option, JPL Tech. Memo. No. TM 391–70 (February 1970).

Vivid motivation for consider covariance analysis is given. The mathematical development is based on the method of normal equations.

[3] Jazwinski, A. H., "Stochastic Processes and Filtering Theory," p. 284. Academic Press, New York, 1970.

Jazwinski's development of the Schmidt consider filter is especially easy to follow. However, two misleading statements appear in the discussion. One of these statements is that the consider filter that does not estimate y is cheaper than the optimal filter that does include y. Both filters require essentially the same amount of storage and do the same amount of computation. *The purpose of the consider filter is to prevent filter divergence, not to save storage or computation.* The second point is that Schmidt's consider filter is not the best estimator of X subject to the constraint y is not estimated. It is optimal only in the sense that it minimizes the covariance trace at the current time, based on the current estimate (which in turn is determined by using nonoptimal prior estimates). A filter that uses different nonoptimal prior estimates might give a better result (smaller covariance trace) for the current estimate.

IX

SRIF Error Analysis of Effects Due to Mismodeled Variables and Incorrect Filter a Priori Statistics

IX.1 Introduction

The accuracy analyses presented in Sections VIII.2 and VIII.3 are conceptually simple and the form of the equations suggests an effortless computer mechanization. Useful and efficient covariance analysis programs have been developed at JPL and elsewhere that are based on these ideas. Success of the SRIF in terms of numerical accuracy, stability, and its various capabilities (cf. Chapters V and VII) motivates an investigation into *its* use for accuracy analysis. Nishimura [1] has married the two methods and developed a square root mechanization of the consider covariance analysis of Section VIII.3. The computer mechanization of Nishimura's algorithms has been the source of numerous JPL navigation analyses.

Instead of algebraically transforming the covariance evaluation algorithms into more accurate and stable square root forms we choose to effectuate the error analysis within the framework of the SRIF. By using the data equation representation for the information array it will be shown that the SRIF error analysis parallels the covariance development of the last chapter. The importance of our independent SRIF accuracy analysis development is that the analyst can better interpret his results and can derive algorithm modifications and/or approximations within the data equation framework. For example, one could analyze the effects of using a "suboptimal" orthogonal transformation in the SRIF in a fashion that is analogous to the way suboptimal gains are evaluated in Kalman filter analyses. A more important reason for the SRIF data equation development is that this method leads directly to efficient, simple algorithms.

The algorithms to be developed here are equivalent to those developed by Nishimura [1], who based his analyses on the generalized Kalman covariance equations (cf. Eqs. (VIII.2.8) and (VIII.2.9)). Although the results are equivalent, Nishimura's computer mechanization involves considerably more storage and computation. At least a part of the computer requirements differential is due to our succinct filter implementation (cf. Appendix VII.B). It is hypothesized that a more fundamental reason for the increased efficiency is that our analyses, which are the subject of this chapter, are carried out within the SRIF data equation framework. This approach to error analysis makes efficient use of the Householder orthogonal transformation, gives new insights, and enhances understanding of square root estimation.

Remark: It is believed that some, if not all, of the improvements discovered via the SRIF analysis can be incorporated into Nishimura's computer implementation.

Use of the names "error analysis" and "accuracy analysis," in an estimation context, is not standard, and these terms, though often used interchangeably, have varied meanings. So that there will be no confusion about our use of these terms we make the following distinctions. *Error analysis* will refer to the process of analyzing the effects of random and systematic errors on the system model and the estimate. Important examples include analyzing errors arising from the use of incorrect *a priori* statistics in the estimation algorithm and the omission of various parameters from the estimation model. The analysis generally involves dealing with a realization of a stochastic process. *Accuracy analysis* will refer to the statistical analysis of estimate errors. Note that error analysis is used to obtain the difference between a realization of a system model and our estimate of the realization; accuracy analysis is used to ascertain the statistical significance of the realization difference. Loosely speaking, error analysis is used to derive algorithms that are used for accuracy analyses.

In this chapter we seek to convey to the reader our opinion that the data equation representation of the SRIF is a source of elegant and efficient accuracy analysis algorithms. The essence of the technique is demonstrated in Section IX.2 where only the effects of incorrect *a priori* statistics are analyzed; included are accuracy analyses of the errors caused by using incorrect *a priori* initial condition, process noise, and measurement noise covariances. In Section IX.3 the analysis is extended to include the effects of unmodeled bias parameters. It is shown that the sensitivity and consider analysis notions, which were introduced in Chapter V and discussed in Chapter VII, are quite useful in this more general setting. The effects of unestimated colored noise parameters are treated in a perfunctory manner because, while the analysis is in principle similar to that described in Sections IX.2 and IX.3, the details would make this chapter too lengthy. Further, we believe (cf. Section VIII.4) that with few exceptions problems with colored noise can be adequately handled by considering them as problems with incorrect a priori statistics. The algorithmic results of the chapter are presented in Appendix IX.B. The algorithm includes, primarily for pedagogical reasons, the effects of unestimated colored noise.

IX.2 Evaluating the Effects of Incorrect Filter a Priori Statistics

Suppose that the model and measurement process have the form

$$X_{j+1} = \Phi_X X_j + G_X w_j \tag{2.1}$$

$$z_j = A_X X_j + \nu_j \tag{2.2}$$

where the dependence of Φ_X, G_X, and A_X on j has been omitted.

Statistics relevant to our analysis are defined as follows.

$[R_X \; z_X]$ Filter information array. The estimate and filter covariance are obtained from the information array via $X_{\text{est}} = R_X^{-1} z_X$ and $R_X^{-1} R_X^{-T}$, respectively. The filter information array is in general independent of any unestimated (consider) parameters. There are exceptions. For example, the Schmidt consider filter of Chapter VIII involved statistics of the consider parameters.

$[R_X^{\text{a}} \; z_X^{\text{a}}]$ "Actual"† information array, based on the true statistics of the problem. Note that

$$X_{\text{est}} = (R_X^{\text{a}})^{-1} z_X^{\text{a}} \quad \text{and} \quad P_X^{\text{a}} = E\left[(X - X_{\text{est}})(X - X_{\text{est}})^{\text{T}}\right] = (R_X^{\text{a}})^{-1}(R_X^{\text{a}})^{-\text{T}}$$

R_w, R_w^{a} Assumed‡ and actual process noise square root information matrices where w is the process noise driving the stochastic parameters of X.

R_ν, R_ν^{a} Assumed and actual measurement noise square root information matrices.

†A superscript "a" will refer to the *actual* or *true* value of a quantity. We do not use a superscript "t" for true because of the possible confusion with the matrix transpose symbol.

‡The reader is cautioned not to associate the "a" superscript with the filter assumed values.

Remark: For simplicity of presentation it is assumed that the process noise and measurement noise have zero means, i.e., $E(w)=0$, $E(\nu)=0$.

Our error analysis algorithm derivation consists of perusing the steps of the filter algorithm and keeping track of the statistics of the error term in the data equation. The method is the same, in principle, as that used to derive the Kalman filter accuracy analysis algorithms.

Remark: In essence *our data equation error analysis technique consists of keeping track of the statistics of the error term in the data equation.*

Initialization

Let the data equation corresponding to the initial estimate be

$$\tilde{z}_X^{\mathrm{a}} = \tilde{R}_X^{\mathrm{a}} X + \tilde{\nu}_X^{\mathrm{a}} \tag{2.3}$$

where $\tilde{\nu}_X^{\mathrm{a}} \in N(0, I)$,[†] and $X_{\mathrm{est}} = (\tilde{R}_X^{\mathrm{a}})^{-1}\tilde{z}_X^{\mathrm{a}}$. The data equation corresponding to the filter assumed statistics is

$$\tilde{z}_X = \tilde{R}_X X + \tilde{\nu}_X \tag{2.4}$$

where in the filter model it is assumed that $\tilde{\nu}_X \in N(0, I)$. To see how $\tilde{\nu}_X$ and $\tilde{\nu}_X^{\mathrm{a}}$ are related, we premultiply Eq. (2.3) by

$$\widetilde{\mathrm{SR}}_X = \tilde{R}_X\left(\tilde{R}_X^{\mathrm{a}}\right)^{-1} \tag{2.5}$$

and find that $\tilde{z}_X = \widetilde{\mathrm{SR}}_X \tilde{z}_X^{\mathrm{a}} = \tilde{R}_X X + \widetilde{\mathrm{SR}}_X \tilde{\nu}_X^{\mathrm{a}}$, i.e., $\tilde{\nu}_X = \widetilde{\mathrm{SR}}_X \tilde{\nu}_X^{\mathrm{a}}$. Thus the incorrect filter assumed estimate covariance is reflected in an incorrect data equation error covariance, and instead of an identity error covariance we have

$$\tilde{P}_{\nu_X} = E(\tilde{\nu}_X \tilde{\nu}_X^{\mathrm{T}}) = \widetilde{\mathrm{SR}}_X(\widetilde{\mathrm{SR}}_X)^{\mathrm{T}} \tag{2.6}$$

Introduction of $\widetilde{\mathrm{SR}}_X$ via (2.5) is not mysterious; it is an obvious way to transform (2.4) to the form of (2.3). Statistical knowledge of $\tilde{\nu}_X$ is maintained through $\widetilde{\mathrm{SR}}_X$; $\widetilde{\mathrm{SR}}_X$ *is a square root of the covariance of* $\tilde{\nu}_X$. When the actual estimate covariance $\tilde{P}_X^{\mathrm{a}}$ is desired it is obtainable from the filter square root information matrix and $\widetilde{\mathrm{SR}}_X$. Using (2.5) we have

$$\tilde{P}_X^{\mathrm{a}} = \left[\tilde{R}_X^{-1}\widetilde{\mathrm{SR}}_X\right]\left[\tilde{R}_X^{-1}\widetilde{\mathrm{SR}}_X\right]^{\mathrm{T}} \tag{2.7}$$

Results (2.5)–(2.7) hold also for the measurement- and time-updated data equations. Updating the data equation error statistics via $\overline{\mathrm{SR}}_X$ is the key to our error analysis approach.[‡]

[†]Recall (cf. page 114) that $N(0, I)$ denotes zero mean and unity covariance.

[‡]We write, as in Chapter VIII, an overbar to represent either ˜ or ˆ.

Data Processing

Let the actual observation be

$$z^{\mathrm{a}} = A_X^{\mathrm{a}} X + \nu^{\mathrm{a}}$$

where $\operatorname{cov}(\nu^{\mathrm{a}}) = E[\nu^{\mathrm{a}}(\nu^{\mathrm{a}})^{\mathrm{T}}] = (R_\nu^{\mathrm{a}})^{-1}(R_\nu^{\mathrm{a}})^{-\mathrm{T}}$. The filter model assumes that $\nu = \nu^{\mathrm{a}}$, and $\operatorname{cov}(\nu) = R_\nu^{-1}R_\nu^{-\mathrm{T}}$. Based on this assumption the observation is scaled by R_ν, i.e.,

$$(R_\nu z^{\mathrm{a}}) = (R_\nu A_X^{\mathrm{a}})X + (R_\nu \nu^{\mathrm{a}})$$

This is Eq. (2.2) with $\nu = R_\nu \nu^{\mathrm{a}}$. The filter model assumes that ν has unit covariance, but this is evidently not the case, i.e.,

$$\operatorname{cov}(\nu) = \left[R_\nu (R_\nu^{\mathrm{a}})^{-1}\right]\left[R_\nu (R_\nu^{\mathrm{a}})^{-1}\right]^{\mathrm{T}} \triangleq (\mathrm{SR}_\nu)(\mathrm{SR}_\nu)^{\mathrm{T}} \tag{2.8}$$

Having ascertained the covariance of the measurement error, we turn our attention to the algorithm for combining the measurement (2.2) and the *a priori* data equation (2.4):

$$\begin{aligned} \tilde{T}_\nu \begin{bmatrix} \tilde{z}_X \\ z \end{bmatrix} &= \tilde{T}_\nu \begin{bmatrix} \tilde{R}_X \\ A_X \end{bmatrix} X + \tilde{T}_\nu \begin{bmatrix} \tilde{\nu}_X \\ \nu \end{bmatrix} \\ \begin{bmatrix} \hat{z}_X \\ * \end{bmatrix} &= \begin{bmatrix} \hat{R}_X \\ 0 \end{bmatrix} X + \begin{bmatrix} \hat{\nu}_X \\ * \end{bmatrix} \end{aligned} \tag{2.9}$$

where $\tilde{T}_\nu$, orthogonal, is a product of elementary Householder transformations designed to triangularize

$$\begin{bmatrix} \tilde{R}_X \\ A_X \end{bmatrix}$$

and the $*$ terms indicate, as is our custom, terms that are of no import to us here.

If $\tilde{T}_\nu$ is partitioned, we can compute $\operatorname{cov}(\hat{\nu}_X)$† from Eq. (2.9), i.e.,

$$\operatorname{cov}(\hat{\nu}_X) = \tilde{T}_{11}(\widetilde{\mathrm{SR}}_X)(\widetilde{\mathrm{SR}}_X)^{\mathrm{T}}\tilde{T}_{11}^{\mathrm{T}} + \tilde{T}_{12}(\mathrm{SR}_\nu)(\mathrm{SR}_\nu)^{\mathrm{T}}\tilde{T}_{12}^{\mathrm{T}}$$

There are several reasons for not using this formula to compute $\operatorname{cov}(\hat{\nu}_X)$. It is computationally more accurate and stable to compute the square root $\widehat{\mathrm{SR}}_X$ directly from $\widetilde{\mathrm{SR}}_X$ and SR_ν without first computing the sum of squares. Further, $\tilde{T}_\nu$ is an implicit quantity (cf. Chapter IV) and is therefore not readily available for partitioning. There is no need to explicitly

†Recall (see p. 178) that $\operatorname{cov}(\hat{\nu}_X) = E[(\nu_X - \hat{\nu}_X)(\nu_X - \hat{\nu}_X)^{\mathrm{T}}]$.

compute $\tilde{T}_\nu$ for partitioning, since $\operatorname{cov}(\hat{\nu}_X)$ can be obtained from

$$\tilde{T}_\nu \left[\begin{array}{c|c} \widetilde{\mathrm{SR}}_X & 0 \\ \hline 0 & \mathrm{SR}_\nu \end{array}\right] = \left[\begin{array}{c|c} \hat{L}_X & \hat{L}_\nu \\ \hline * & * \end{array}\right] \tag{2.10}$$

$$\operatorname{cov}(\hat{\nu}_X) = \hat{L}_X \hat{L}_X^{\mathrm{T}} + \hat{L}_\nu \hat{L}_\nu^{\mathrm{T}} = \widehat{\mathrm{SR}}_X \widehat{\mathrm{SR}}_X^{\mathrm{T}}$$

Further, from this "sum of squares" expression it is easy to conclude that one method of computing $\widehat{\mathrm{SR}}_X$ is to choose an orthogonal transformation $\hat{T}_L$ such that

$$[0 \quad \widehat{SR}_X] = [\hat{L}_X \quad \hat{L}_\nu]\hat{T}_L \tag{2.11}$$

where $\widehat{\mathrm{SR}}_X$ is upper triangular.

Remark: Computation of (2.11) can be realized using algorithms that are simple modifications of those described in Chapter IV.

An Aside: It is easy to extend this analysis to include the effect of coefficient mismodeling. To illustrate the ideas that are involved we consider the effect of replacing

$$z = Ax + \nu, \qquad \text{where} \quad \nu \in N(0, I) \tag{2.12}$$

by

$$z = \tilde{A}x + \tilde{\nu} \tag{2.13}$$

where $\tilde{A}$ is an approximation or simplification of A, and for filter computation purposes it is assumed that $\tilde{\nu} \in N(0, I)$. We assume $A(m, n)$ with $m > n$. If $\tilde{R}$, ΔR, $\tilde{z}$, and ν_A are defined by

$$\tilde{T}[\tilde{A} \quad \Delta A \quad z \quad \nu] = \begin{bmatrix} \tilde{R} & \Delta R & \tilde{z} & \nu_A \\ 0 & * & * & * \end{bmatrix} \tag{2.14}$$

where $\tilde{T}$, orthogonal, is chosen to triangularize $\tilde{A}$ and $\Delta A = A - \tilde{A}$, then

$$\tilde{x} = \tilde{R}^{-1}\tilde{z} \qquad \text{(computed estimate)} \tag{2.15a}$$

$$\tilde{P}_{\mathrm{c}} = \tilde{R}^{-1}\tilde{R}^{-\mathrm{T}} \qquad \text{(computed covariance)} \tag{2.15b}$$

$$x - \tilde{x} = \tilde{R}^{-1}\,\Delta R x - \tilde{R}^{-1}\nu_A \qquad \text{(estimate error)} \tag{2.15c}$$

Assuming that x and ν are statistically independent, we obtain the actual covariance from (2.15c)

$$\operatorname{cov}(\tilde{x}) = \tilde{P}_{\mathrm{c}} + \tilde{S}_A E(xx^{\mathrm{T}})\tilde{S}_A^{\mathrm{T}} \tag{2.16}$$

where

$$\tilde{S}_A = -\tilde{R}^{-1}\,\Delta R \qquad \text{(coefficient mismodeling sensitivity)}$$

and $E(xx^{\mathrm{T}})$ is the *a priori* covariance of an initial zero estimate of x. We note that (2.15c) and (2.16) generalize the sensitivity results of Section VII.2 (cf. Eqs. (VII.2.15) and (VII.2.18)). Result (2.16) can also be used to represent the effect of using incorrect estimate *a priori* statistics in the filter. To see this, note that the *a priori* statistics, represented in data equation form, are included in Eqs. (2.12) and (2.13). ■

Propagation in Time

The SRIF propagation algorithm of Chapter V is

$$\tilde{T}_w \begin{bmatrix} z_w \\ \hat{z}_X \end{bmatrix} = \tilde{T}_w \begin{bmatrix} R_w & 0 \\ -R_X^{\mathrm{d}} G_X & R_X^{\mathrm{d}} \end{bmatrix} \begin{bmatrix} w_j \\ X_{j+1} \end{bmatrix} + \tilde{T}_w \begin{bmatrix} \nu_w \\ \hat{\nu}_X \end{bmatrix}$$

$$\begin{bmatrix} * \\ \tilde{z}_X \end{bmatrix} = \begin{bmatrix} * & * \\ 0 & \tilde{R}_X \end{bmatrix} \begin{bmatrix} w_j \\ X_{j+1} \end{bmatrix} + \begin{bmatrix} * \\ \tilde{\nu}_X \end{bmatrix} \tag{2.17}$$

where $R_X^{\mathrm{d}} = \hat{R}_X \Phi_X^{-1}$, and $\tilde{T}_w$ is a product of elementary Householder transformations chosen to partially triangularize the augmented state square root information matrix.

The covariance of $\tilde{\nu}_X$ is calculated in the same way as before (cf. Eqs. (2.10) and (2.11)), i.e.,

$$\tilde{T}_w \begin{bmatrix} \mathrm{SR}_w & 0 \\ 0 & \widehat{\mathrm{SR}}_X \end{bmatrix} = \begin{bmatrix} * & * \\ \tilde{L}_w & \tilde{L}_X \end{bmatrix} \tag{2.18}$$

$$[0 \quad \widetilde{SR}_X] = [\tilde{L}_w \quad \tilde{L}_X]\tilde{T}_L \tag{2.19}$$

where

$$\mathrm{SR}_w = R_w (R_w^{\mathrm{a}})^{-1} \tag{2.20}$$

is the square root covariance of ν_w.

The significance of representations (2.10) and (2.18) is that we can mechanize the SR_X computations by merely addending additional matrix columns to the SRIF tableau. The combined algorithm is succinctly expressed as follows.

Data Inclusion

$$\hat{T}_\nu \begin{bmatrix} \overbrace{\tilde{R}_X}^{N_X} & \overbrace{\tilde{z}_X}^{1} & \overbrace{\widetilde{\mathrm{SR}}_X}^{N_X} & \overbrace{0}^{m} \\ A_X & z & 0 & \mathrm{SR}_\nu \end{bmatrix} \begin{matrix} \}N_X \\ \}m \end{matrix}$$

$$= \begin{bmatrix} \hat{R}_X & \hat{z}_X & \hat{L}_X & \hat{L}_\nu \\ 0 & e & * & * \end{bmatrix} \begin{matrix} \}N_X \\ \}m \end{matrix} \tag{2.21}$$

$$[0 \quad \widehat{\mathrm{SR}}_X] = \left[\hat{L}_X \quad \hat{L}_\nu\right]\hat{T}_L \tag{2.11}$$

Propagation

$$\tilde{T}_w \begin{bmatrix} \overbrace{R_w}^{N_w} & \overbrace{0}^{N_X} & \overbrace{z_w}^{1} & \overbrace{\mathrm{SR}_w}^{N_w} & \overbrace{0}^{N_X} \\ -R_X^{\mathrm{d}}G_X & R_X^{\mathrm{d}} & \hat{z}_X & 0 & \widehat{\mathrm{SR}}_X \end{bmatrix} \begin{matrix} \}N_w \\ \}N_X \end{matrix}$$

$$= \begin{bmatrix} * & * & * & * & * \\ 0 & \tilde{R}_X & \tilde{z}_X & \tilde{L}_w & \tilde{L}_X \end{bmatrix} \begin{matrix} \}N_w \\ \}N_X \end{matrix} \tag{2.22}$$

$$[0 \quad \widetilde{\mathrm{SR}}_X] = \left[\tilde{L}_w \quad \tilde{L}_X\right]\tilde{T}_L \tag{2.19}$$

Remarks: (1) When SR_w is diagonal (as is usually the case in practice), one can arrange computer implementation so that (2.22) can be cycled through N_w times, with each cycle corresponding to a problem with $N_w = 1$. Such an implementation reduces computer storage requirements. Computations too are reduced, but the reduction is only modest. Arranging the measurement update (2.21) in an analogous one-component-at-a-time fashion reduces storage; however, computational requirements in this case are significantly increased. Thus one must trade off storage savings against increased computation. Experience suggests avoiding excessive storage requirements and therefore limiting the size of the measurement vectors.

(2) Note that one could, with additional column augmentation (cf. the aside on page 188), include the effects of transition matrix model errors.

(3) Observe that with few exceptions, dependence on the index j has

been omitted. Our practice is to derive equations for ^ quantities in terms of ~ quantities, and conversely. It is understood that time is advancing.

IX.3 Considering the Effects of Unmodeled Biases

In this section we propose a variation of the P_X^{a} performance calculation that is chosen to highlight the similarities of the estimated and consider bias parameters. Our goal is to delay deciding which parameters are to be estimated until after the data has been processed. A less important but pragmatic motive is that partitioning R_X eliminates a block of zeros, R_{yp} and R_{yx}, and at the same time puts the results in a form that is compatible with the filter mechanization of Chapter VII. The main result of this section is the conclusion that consider bias analysis can be decided *a posteriori* and implemented by partitioning the y bias parameter vector into estimated and consider segments. Such a decomposition is already used (cf. Chapters V and VII) for SRIF analyses that do not involve incorrect *a priori* statistics. The significance of the present result is that it shows that bias parameter sensitivity can be computed the same way, whether or not incorrect covariance *a priori* are used for the estimated parameters, process noise, and measurement noise.

The model used to construct our filter estimate is the one described in Chapter VII, Eqs. (VII.1.1) and (VII.1.2), and we let

$$X = \begin{cases} p & \text{Exponentially correlated process noise, } N_p \text{ in number} \\ x & \text{States that are time varying, but do not explicitly depend} \\ & \text{on white process noise, } N_x \text{ in number} \\ y & \text{Bias parameters, } N_y \text{ in number} \end{cases}$$

Remarks: (1) The X model is adequate for a rather wide range of applications, especially since the assumption of a diagonal process noise dynamics matrix M in Eq. (VII.1.2) is not an essential requirement. Note that the presence of process noise terms in the x model equation (cf. Eq. (VII.1.1)) is not ruled out because one can use p's that are white noise, with corresponding m's that are zero.

(2) The SRIF accuracy analysis algorithm developments in this section involve formidable arrays with lengthy matrix arguments. Our purpose is mainly to demonstrate the role played by the filter orthogonal transformations, and the utility of using augmented filter arrays.

Our derivation is rather pedestrian, and consists in the main of substituting the notation of Chapter VII into the results of Section IX.2. To begin, we use the notation of Section VII.3 and write data equations for p, x, and y:

$$N_p + N_x \left\{ \left[\overbrace{S_p^{\mathrm{a}}}^{N_p} \quad \overbrace{S_x^{\mathrm{a}}}^{N_x} \quad \overbrace{S_y^{\mathrm{a}}}^{N_y} \right] \right. \begin{bmatrix} p \\ x \\ y \end{bmatrix} = S_z^{\mathrm{a}} - S_\nu^{\mathrm{a}}, \quad R_y^{\mathrm{a}} y = z_y^{\mathrm{a}} - \nu_y^{\mathrm{a}} \tag{3.1}$$

where S_ν^{a} and ν_y^{a} are zero mean and have unit covariances. The filter is similarly initialized, except without the superscript "a." Recall from Chapter VII, pages 143–144, that the S notation is related to the R's, z's, and ν's, i.e.,

$$S_p = \begin{bmatrix} R_p \\ R_{xp} \end{bmatrix}, \quad S_x = \begin{bmatrix} R_{px} \\ R_x \end{bmatrix}, \quad S_y = \begin{bmatrix} R_{py} \\ R_{xy} \end{bmatrix}, \quad S_z = \begin{bmatrix} z_p \\ z_x \end{bmatrix} \tag{3.2}$$

and

$$S = [S_p \quad S_x \quad S_y \quad S_z]$$

The z's and ν's, with and without subscripts, are related to each other via SR_X (cf. Eq. (2.5), et seq.):

$$S_z = \mathrm{SR}_X S_z^{\mathrm{a}} \tag{3.3}$$

$$S_\nu = \mathrm{SR}_X S_\nu^{\mathrm{a}} \tag{3.4}$$

As noted earlier, (3.4) is the key to relating the actual and assumed square root information matrices

$$\mathrm{SR}_X \triangleq \begin{bmatrix} \mathrm{sr}_p & \mathrm{sr}_{px} & \mathrm{sr}_{py} \\ \mathrm{sr}_{xp} & \mathrm{sr}_x & \mathrm{sr}_{xy} \\ 0 & 0 & \mathrm{sr}_y \end{bmatrix} = R_X (R_X^{\mathrm{a}})^{-1} \tag{3.5}$$

This equation is used to initialize SR_X and is used later to construct

$$P_X^{\mathrm{a}} \quad \left(= (R_X^{\mathrm{a}})^{-1} (R_X^{\mathrm{a}})^{-\mathrm{T}}\right) \qquad \text{(cf. Eq. (2.7))}$$

Remarks: (1) Lower case sr's are used to avoid notational confusions (cf. Eqs. (3.6) and (3.7) that follow). There we have

$$\mathrm{SR}_p = \begin{bmatrix} \mathrm{sr}_p \\ \mathrm{sr}_{xp} \end{bmatrix}, \quad \mathrm{SR}_x = \begin{bmatrix} \mathrm{sr}_{px} \\ \mathrm{sr}_x \end{bmatrix}, \quad \mathrm{SR}_y, \quad \text{and} \quad \mathrm{sr}_y$$

These variables all relate to matrix square roots.

(2) If there is no initial y correlation, then one need not include y covariance *a priori* information. In this case sr_y is initially zero. There is no concern about the singularity of SR_X since it does not have to be inverted.

(3) Equation (3.3) comes about because S_z^{a} must correspond to the same estimate as S_z.

Since we intend to partition the results (2.21), (2.11), (2.22), and (2.19) we introduce the two matrix tableaux

$$\begin{bmatrix} R_p & R_{px} & R_{py} & z_p & \mathrm{sr}_p & \mathrm{sr}_{px} & \mathrm{sr}_{py} \\ R_{xp} & R_x & R_{xy} & z_x & \mathrm{sr}_{xp} & \mathrm{sr}_x & \mathrm{sr}_{xy} \end{bmatrix}$$
$$= \begin{bmatrix} S_p & S_x & S_y & S_z & \mathrm{SR}_p & \mathrm{SR}_x & \mathrm{SR}_y \end{bmatrix} \tag{3.6}$$

$$[R_y \quad z_y \quad \mathrm{sr}_y] \tag{3.7}$$

They correspond to the p–x and y filter information arrays, augmented with appropriate square root data equation error terms.

The data inclusion algorithm corresponding to (2.21) and (2.11) is derived by modifying data equation (2.9). We combine the measurement (VII.2.3) with the filter data equations, as in Eqs. (VII.2.6) and (VII.2.7), and obtain[†]

$$\hat{T}_{px}\begin{bmatrix} \tilde{S}_p & \tilde{S}_x & \tilde{S}_y \\ A_p & A_x & A_y \end{bmatrix}\begin{bmatrix} p \\ x \\ y \end{bmatrix} = \hat{T}_{px}\begin{bmatrix} \tilde{S}_z \\ z \end{bmatrix} - T_{px}\begin{bmatrix} \tilde{S}_\nu \\ \nu \end{bmatrix}$$
$$\begin{bmatrix} \hat{S}_p & \hat{S}_x & \hat{S}_y \\ 0 & 0 & \hat{A}_y \end{bmatrix}\begin{bmatrix} p \\ x \\ y \end{bmatrix} = \begin{bmatrix} \hat{S}_z \\ \hat{z} \end{bmatrix} - \begin{bmatrix} \hat{S}_\nu \\ \hat{\nu} \end{bmatrix} \tag{3.8}$$

where $\hat{T}_{px}$, which is independent of y, is chosen so that

$$\hat{T}_{px}\begin{bmatrix} \tilde{S}_p & \tilde{S}_x \\ A_p & A_x \end{bmatrix} = \begin{bmatrix} \hat{S}_p & \hat{S}_x \\ 0 & 0 \end{bmatrix}$$

[†]Note that

$$S_\nu = \begin{bmatrix} \nu_p \\ \nu_x \end{bmatrix}$$

and $[\hat{S}_p \ \hat{S}_x]$ is an upper triangular matrix. The reduced measurement $\hat{A}_y y = \hat{z} - \hat{\nu}$ from Eq. (3.8) is now combined with the y *a priori*, yielding

$$\hat{T}_y \begin{bmatrix} \tilde{R}_y \\ \hat{A}_y \end{bmatrix} y = \hat{T}_y \begin{bmatrix} \tilde{z}_y \\ \hat{z} \end{bmatrix} - \hat{T}_y \begin{bmatrix} \tilde{\nu}_y \\ \hat{\nu} \end{bmatrix}$$
$$\begin{bmatrix} \hat{R}_y \\ 0 \end{bmatrix} y = \begin{bmatrix} \hat{z}_y \\ e \end{bmatrix} - \begin{bmatrix} \hat{\nu}_y \\ * \end{bmatrix} \tag{3.9}$$

To see how $\widetilde{\mathrm{SR}}_X$ is transformed by the orthogonal transformations we partition the $\hat{T}$ matrices and write the $\wedge$ error terms

$$\begin{aligned} \hat{S}_\nu &= \hat{T}_{px}(1,1)\tilde{S}_\nu + \hat{T}_{px}(1,2)\nu \\ \hat{\nu} &= \hat{T}_{px}(2,1)\tilde{S}_\nu + \hat{T}_{px}(2,2)\nu \\ \hat{\nu}_y &= \hat{T}_y(1,1)\hat{\nu}_y + \hat{T}_y(1,2)\hat{\nu} \end{aligned} \tag{3.10}$$

We will form an updated SR_X based on these expressions. The technique is the same as that used to obtain (2.21) and (2.11), and the results to be derived via Eqs. (3.13)–(3.16) are the data inclusion algorithms

$$\hat{T}_{px} \begin{bmatrix} \tilde{S}_p & \tilde{S}_x & \tilde{S}_y & \tilde{S}_z & \widetilde{\mathrm{SR}}_p & \widetilde{\mathrm{SR}}_x & \widetilde{\mathrm{SR}}_y & 0 \\ A_p & A_x & A_y & z & 0 & 0 & 0 & \mathrm{SR}_\nu \end{bmatrix}$$
$$= \begin{bmatrix} \hat{S}_p & \hat{S}_x & \hat{S}_y & \hat{S}_z & \hat{L}_p & \hat{L}_x & \hat{L}_y & \hat{L}_\nu \\ 0 & 0 & \hat{A}_y & \hat{z} & \hat{M}_p & \hat{M}_x & \hat{M}_y & \hat{M}_\nu \end{bmatrix} \tag{3.11}$$

and

$$\tilde{T}_y \begin{bmatrix} \tilde{R}_y & \tilde{z}_y & 0 & 0 & \widetilde{\mathrm{sr}}_y & 0 \\ \hat{A}_y & \hat{z} & \hat{M}_p & \hat{M}_x & \hat{M}_y & \hat{M}_\nu \end{bmatrix}$$
$$= \begin{bmatrix} \hat{R}_y & \hat{z}_y & H_p & H_x & H_y & H_\nu \\ 0 & e & * & * & * & * \end{bmatrix} \tag{3.12}$$

To see how the $\hat{L}$'s and H's are used to construct $\widehat{\mathrm{SR}}_X$, note that the key formulae (3.10) can be written as

$$\begin{bmatrix} \hat{S}_\nu \\ \hat{\nu}_y \end{bmatrix} = \begin{bmatrix} \hat{T}_{px}(1,1) & 0 \\ \hat{T}_y(1,2)\hat{T}_{px}(2,1) & \hat{T}_y(1,1) \end{bmatrix} \begin{bmatrix} \tilde{S}_\nu \\ \tilde{\nu}_y \end{bmatrix} + \begin{bmatrix} \hat{T}_{px}(1,2) \\ \hat{T}_y(1,2)T_{px}(2,2) \end{bmatrix} \nu \tag{3.13}$$

From this equation, however, one can see by inspection† that the following array (3.14a) is a rectangular square root of $\operatorname{cov}[(\hat{S}_\nu^{\mathrm{T}}\ \hat{\nu}_y^{\mathrm{T}})^{\mathrm{T}}]$:

$$\left[\begin{bmatrix} \hat{T}_{px}(1,1) & 0 \\ \hat{T}_y(1,2)\hat{T}_{px}(2,1) & \hat{T}_y(1,1) \end{bmatrix} \begin{bmatrix} \widetilde{\mathrm{SR}}_p & \widetilde{\mathrm{SR}}_x & \widetilde{\mathrm{SR}}_y \\ 0 & 0 & \widetilde{\mathrm{sr}}_y \end{bmatrix} \;\middle|\; \begin{matrix} \hat{T}_{px}(1,2)\mathrm{SR}_\nu \\ \hat{T}_y(1,2)\hat{T}_{px}(2,2)\mathrm{SR}_\nu \end{matrix}\right] \tag{3.14a}$$

$$= \left[\begin{array}{cccc|c} \hat{L}_p & \hat{L}_x & & \hat{L}_y & \hat{L}_\nu \\ \hat{T}_y(1,2)\hat{M}_p & \hat{T}_y(1,2)\hat{M}_x & & \hat{T}_y(1,2)\hat{M}_y + \hat{T}_y(1,1)\widetilde{\mathrm{sr}}_y & \hat{T}_y(1,2)\hat{M}_\nu \end{array}\right] \tag{3.14b}$$

$$= \begin{bmatrix} \hat{L}_p & \hat{L}_x & \hat{L}_y & \hat{L}_\nu \\ H_p & H_x & H_y & H_\nu \end{bmatrix} \tag{3.14c}$$

Equations (3.14b) and (3.14c) are obtained by algebraically expanding (3.14a) and comparing the results with Eqs. (3.11) and (3.12). To convert the rectangular square root (3.14c) to a true‡ square root we apply an orthogonal transformation $\hat{T}_\nu$ to zero out the excess columns. The resultant measurement-updated square root error terms are given by

$$\left[\begin{array}{c|ccc} \overbrace{0}^{m} & \widehat{\mathrm{SR}}_p & \widehat{\mathrm{SR}}_x & \widehat{\mathrm{SR}}_y \\ & 0 & 0 & \widehat{\mathrm{sr}}_y \end{array}\right] = \begin{bmatrix} \hat{L}_p & \hat{L}_x & \hat{L}_y & \hat{L}_\nu \\ H_p & H_x & H_y & H_\nu \end{bmatrix} \hat{T}_\nu \tag{3.15}$$

Remark: Note that the SR related matrices can require a considerable investment of temporary computer memory when the number of measurements m is even moderately large. The memory requirement problem can be alleviated by processing the measurements in smaller batches. Of course smaller data batches increases the computational cost and a compromise between computer storage and computational cost may be necessary. For example, if $N_p + N_x + N_y > 50$, then it would be wise to restrict m to, say, $m \leqslant 10$.

Having completed the derivation of the data inclusion algorithm we turn now to the problem of propagating the square root error covariance matrix when the model (VII.1.1) is used. The explicitness of the model makes it feasible to replace the general SRIF propagation algorithm (2.17) by the

†If $\gamma = A\alpha + B\beta$, with a and β independent random vectors having associated covariance matrices P_α and P_β, respectively; then $[AP_\alpha^{1/2}\quad BP_\beta^{1/2}]$ is a rectangular square root of P_γ. To verify that this is true, compute $P_\gamma^{1/2}P_\gamma^{\mathrm{T}/2}$.

‡Recall that our matrix square root definition (cf. page 37) involved square matrices.

colored noise propagation (VII.3.4). Writing the filter propagation algorithm in data equation terms will lead straightaway to expressions for $\widetilde{\mathrm{SR}}_x$:

$$\tilde{T}_p \begin{bmatrix} -R_w M & R_w & 0 & 0 \\ \hat{S}_p - S_x^{\mathrm{d}} V_p & 0 & S_x^{\mathrm{d}} & \hat{S}_y - S_x^{\mathrm{d}} V_y \end{bmatrix} \begin{bmatrix} p_j \\ p_{j+1} \\ x_{j+1} \\ y \end{bmatrix}$$

$$= \tilde{T}_p \begin{bmatrix} z_w \\ \hat{S}_z \end{bmatrix} - \tilde{T}_p \begin{bmatrix} \nu_w \\ \hat{S}_\nu \end{bmatrix} \tag{3.16}$$

$$\begin{bmatrix} * & * & * & * \\ 0 & \tilde{S}_p & \tilde{S}_x & \tilde{S}_y \end{bmatrix} \begin{bmatrix} p_j \\ p_{j+1} \\ x_{j+1} \\ y \end{bmatrix} = \begin{bmatrix} * \\ \tilde{S}_z \end{bmatrix} - \begin{bmatrix} * \\ \tilde{S}_\nu \end{bmatrix}$$

where $S_x^{\mathrm{d}} = \hat{S}_x U^{-1}$, and the $*$ terms indicate that they play no role in what follows. The $\tilde{}$ terms are the result of the propagation (cf. Eq. (VI.3.6)). Carrying through steps analogous to (2.18) we write

$$\tilde{T}_p \begin{bmatrix} \mathrm{SR}_w & 0 & 0 & 0 \\ 0 & \widehat{\mathrm{SR}}_p & \widehat{\mathrm{SR}}_x & \widehat{\mathrm{SR}}_y \end{bmatrix} = \begin{bmatrix} * & * & * & * \\ \tilde{L}_w & \tilde{L}_p & \tilde{L}_x & \tilde{L}_y \end{bmatrix} \tag{3.17}$$

and note that

$$\operatorname{cov}(\tilde{S}_\nu) = \tilde{L}_w \tilde{L}_w^{\mathrm{T}} + \tilde{L}_p \tilde{L}_p^{\mathrm{T}} + \tilde{L}_x \tilde{L}_x^{\mathrm{T}} + \tilde{L}_y \tilde{L}_y^{\mathrm{T}} \tag{3.18}$$

Recall that our plan is to update the covariance of the data equation error $[S_\nu^{\mathrm{T}}\ \nu_y^{\mathrm{T}}]^{\mathrm{T}}$. The effect of $\tilde{T}_p$ on S_ν is given in (3.16) and its updated covariance is given by (3.18). Note that the y data equation error ν_y is time independent and therefore sr_y is unchanged. The change in S_ν does, however, affect the correlation between S_ν and ν_y. This correlation is related to our square root factorization by expanding $P = SS^{\mathrm{T}}$, with S upper triangular, i.e.,

$$P_{S_\nu \nu_y} = \begin{bmatrix} P_{\nu_p \nu_y} \\ P_{\nu_x \nu_y} \end{bmatrix} = \mathrm{SR}_y \mathrm{sr}_y^{\mathrm{T}} \tag{3.19}$$

Using this expression and the rightmost columns of Eq. (3.17), we conclude that

$$\widetilde{\mathrm{SR}}_y = \tilde{L}_y \tag{3.20}$$

and a further comparison of the expansion of $\mathrm{cov}[(\tilde{S}_v^{\mathrm{T}}\ \nu_y^{\mathrm{T}})^{\mathrm{T}}]$ leads to

$$[0\ \ \widetilde{\mathrm{SR}}_p\ \ \widetilde{\mathrm{SR}}_x] = \left[\tilde{L}_w\ \ \tilde{L}_p\ \ \tilde{L}_x\right]\tilde{T}_{\mathrm{sr}} \tag{3.21}$$

where the orthogonal transformation $\tilde{T}_{\mathrm{sr}}$ is chosen so that $\widetilde{\mathrm{SR}}_p$ is an upper triangular matrix.

The reason for arranging the $\tilde{L}$ algorithm in the form (3.17) is that we plan to operate on an augmented array with the filter transformation $\tilde{T}_p$, which is written as a product of elementary Householder transformations $\tilde{T}_p = \tilde{T}(N_p) \cdots \tilde{T}(1)$, where $\tilde{T}(k)$ only alters the elements in row k and the bottom $N_p + N_x$ rows. In Appendix IX.A it is shown that the $\tilde{L}$ array can be generated via

$$1\{\begin{bmatrix} * \text{---} * \\ L_k \end{bmatrix} = \tilde{T}_k \begin{bmatrix} \overbrace{\mathrm{SR}_w(k)}^{1} & 0 \\ 0 & L_{k-1} \end{bmatrix}\}1 \qquad k = 1, \ldots, N_p \tag{3.22†}$$

with

$$L_0 = [\widehat{\mathrm{SR}}_p\ \ \widehat{\mathrm{SR}}_x\ \ \widehat{\mathrm{SR}}_y], \qquad \text{and} \qquad L_{N_p} = \left[\tilde{L}_w\ \ \tilde{L}_p\ \ \tilde{L}_x\ \ \tilde{L}_y\right]$$

and where $\tilde{T}_k$ is that part of $\tilde{T}(k)$ that corresponds to the filter mechanization. For reference purposes we include the augmented array algorithm. The algorithm includes the efficient mechanization of the filter propagation algorithm that is implemented in Appendix VII.B. We avoid introducing new notation by using the FORTRAN technique of writing over old variables. A perhaps more pragmatic reason for using such notation is that it facilitates computer implementation. The algorithm follows:

First set

$$\tilde{S}_x = \hat{S}_x U^{-1}, \qquad \tilde{S}_y = \hat{S}_y - \tilde{S}_x V_y$$

$$\tilde{L}_0 = [\widehat{\mathrm{SR}}_p\ \ \widehat{\mathrm{SR}}_x\ \ \widehat{\mathrm{SR}}_y], \qquad \left[\tilde{S}_p\ \ \tilde{S}_z\right] = \left[\tilde{S}_p\ \ \hat{S}_z\right]$$

Then for $k = 1, \ldots, N_p$ recursively evaluate Eqs. (3.23)–(3.24)

$$\tilde{S}_p(1) := \tilde{S}_p(1) - \tilde{S}_x V_p(k) \tag{3.23}$$

†For our purposes it suffices to assume that SR_w is diagonal. The more general case of an upper triangular SR_w matrix would result in a different, but essentially similar, algorithm.

$$\tilde{T}_k \left[\begin{array}{cc|cccccc} \overbrace{-R_w(k)m(k)}^{1} & \overbrace{0}^{N_p-1} & \overbrace{R_w(k)}^{1} & \overbrace{0}^{N_x} & \overbrace{0}^{N_y} & \overbrace{z_w(k)}^{1} & \overbrace{\mathrm{SR}_w(k)}^{1} & \overbrace{0}^{N_p+N_x+k} \\ \tilde{S}_p & & 0 & \tilde{S}_x & \tilde{S}_y & \tilde{S}_z & 0 & \tilde{L}_{k-1} \end{array}\right] \begin{matrix} \}1 \\ \}N_p + N_x \end{matrix}$$

$$:= \left[\begin{array}{c|ccccc} \overbrace{*}^{1} & * & & & & * \\ 0 & \tilde{S}_p & \tilde{S}_x & \tilde{S}_y & \tilde{S}_z & \tilde{L}_k \end{array}\right] \tag{3.24}$$

$\tilde{T}_k$ is an elementary Householder transformation chosen to zero the first column (except, of course, for the first element), and $S_p(1)$ and $V_p(k)$ are the first and kth columns of their respective matrices. The $*$ terms of (3.24) are not relevant to our current investigation, although they are of use in smoothing and in smoothing error analyses.

Remark: Note that algorithm (3.23)–(3.24) reduces computer storage requirements considerably over what they might be if Eq. (VII.3.6) and (3.17) had been implemented. Indeed, computer storage is reduced from $2(N_p + N_x)[2(2N_p + 2N_x + N_y) + 1]$ (using Eqs. (VII.3.6) and (IX.3.17)) to $(N_p + N_x + 1)[3N_p + 2N_x + N_y + 1]$ (using Eq. (3.24)). Even this last count is conservative, since many of the zero terms can be omitted. A set of typical dimensions (for spacecraft trajectory analyses) for these parameters might be $N_p = 10$, $N_x = 6$, $N_y = 15$. Using the first storage arrangement would require an array with 3040 single precision elements while the array used by (3.24) requires only 782 single precision elements.

Now that the augmented information array recursions have been defined we turn to the problem of constructing consider covariances. The data equation for the p and x variables is

$$[S_p \quad S_x]\begin{bmatrix} p \\ x \end{bmatrix} + S_y y = S_z - S_v \tag{3.25}$$

Let us define

$$\begin{bmatrix} p \\ x \end{bmatrix}_{\mathrm{bf}} = [S_p \quad S_x]^{-1} S_z \tag{3.26}$$

where the subscript "bf" represents bias free, i.e., independent of y. Further, y is partitioned as

$$y = \begin{bmatrix} y_1 \\ y_2 \end{bmatrix} \tag{3.27}$$

where the y_1 variables are estimated and the y_2 variables are considered (i.e., they are omitted from the filter). Consistent with this parameter partition we set

$$\text{Sen} = [\text{Sen}_1 \quad \text{Sen}_2] = -[S_p \quad S_x]^{-1}[S_{y_1} \quad S_{y_2}] \tag{3.28}$$

In terms of the notation just introduced one can write

$$\begin{bmatrix} p \\ x \end{bmatrix} = \begin{bmatrix} p \\ x \end{bmatrix}_{\text{bf}} + \text{Sen}_1 y_1 + \text{Sen}_2 y_2 - [S_p \; S_x]^{-1} S_\nu \tag{3.29}$$

Remark: In Chapter VII the term $[{}^p_x]_{\text{bf}}$ was called the "computed estimate." More generally, when we speak of computed estimate, we mean

$$\begin{bmatrix} p^c \\ x^c \end{bmatrix} = \begin{bmatrix} p \\ x \end{bmatrix}_{\text{bf}} + \text{Sen}_1 y_1^c \tag{3.30}$$

where y_1^c is the filter estimate of the model that omits y_2. The notation in Chapter VII thus applied only to the the case where there were no estimated y parameters and to the case where y_1 was regarded as part of x.

The columns of Sen are doubly useful. Perusal of these columns often suggests which biases can most affect the estimate. The second utilitarian function of the sensitivity matrix is that one can use it to determine incremental covariance errors due to considering individual biases. Thus the vector $\text{Sen}(\cdot, j)\sigma_y(j)$† can be thought of as a 1σ perturbation of the estimate due to uncertainty in the parameter $y(j)$; and the variance in the ith component of the estimate due to all of these y uncertainties is $\Sigma_j[\text{Sen}(i, j)\sigma_y(j)]^2$. This last statement is only of qualitative value when the components of y are correlated.

To obtain the consider or true covariance of the filter estimate the data equations for p, x, and y_1 are written as

$$\begin{bmatrix} S_p & S_x & S_{y_1} \\ 0 & 0 & R_{y_1} \end{bmatrix} \begin{bmatrix} p \\ x \\ \hline y_1 \end{bmatrix} = \begin{bmatrix} S_z \\ z_{y_1} \end{bmatrix} - \begin{bmatrix} S_{y_2} \\ R_{y_1 y_2} \end{bmatrix} y_2 - \begin{bmatrix} S_\nu \\ \nu_{y_1} \end{bmatrix} \tag{3.31}$$

†$\text{Sen}(\cdot, j)$ is the jth column of the Sen matrix.

where the y_1 and y_2 terms correspond to the y partitioning

$$R_y = \begin{bmatrix} R_{y_1} & R_{y_1 y_2} \\ 0 & R_{y_2} \end{bmatrix}; \qquad z_y = \begin{bmatrix} z_{y_1} \\ z_{y_2} \end{bmatrix}; \qquad \nu_y = \begin{bmatrix} \nu_{y_1} \\ \nu_{y_2} \end{bmatrix}$$

and

$$\mathrm{sr}_y = \begin{bmatrix} \mathrm{sr}_{y_1} & \mathrm{sr}_{y_1 y_2} \\ 0 & \mathrm{sr}_{y_2} \end{bmatrix}$$

Recall that the filter estimate of p, x, and y_1 satisfies

$$\begin{bmatrix} S_p & S_x & S_{y_1} \\ 0 & 0 & R_{y_1} \end{bmatrix} \begin{bmatrix} \hat{p} \\ \hat{x} \\ \hline \hat{y}_1 \end{bmatrix} = \begin{bmatrix} S_z \\ z_{y_1} \end{bmatrix} \tag{3.32}$$

i.e., (3.31) is used, but with $\hat{y}_2 = 0$ and $\hat{\nu} = 0$. Note that $\hat{p}$, $\hat{x}$, and $\hat{y}_1$ are given by

$$\hat{y}_1 = R_{y_1}^{-1} z_{y_1}, \qquad \begin{bmatrix} \hat{p} \\ \hat{x} \end{bmatrix} = \begin{bmatrix} p \\ x \end{bmatrix}_{\mathrm{bf}} + \mathrm{Sen}_1\, \hat{y}_1 \tag{3.33}$$

Subtracting (3.31) from (3.32) we obtain the vector error equation for p, x, and y_1:

$$\begin{bmatrix} S_p & S_x & S_{y_1} \\ 0 & 0 & R_{y_1} \end{bmatrix} \begin{bmatrix} p - \hat{p} \\ x - \hat{x} \\ \hline y_1 - \hat{y}_1 \end{bmatrix} = \eta = \begin{bmatrix} \eta_{px} \\ \eta_{y_1} \end{bmatrix}$$

where

$$\eta = -\begin{bmatrix} S_{y_2} \\ R_{y_1 y_2} \end{bmatrix} y_2 - \begin{bmatrix} S_\nu \\ \nu_{y_1} \end{bmatrix}$$

Note that the filter model assumes that η has unit covariance, and that a filter covariance square root is

$$\begin{bmatrix} P_{(px)} & P_{(px)y_1} \\ P_{y_1(px)} & P_{y_1} \end{bmatrix}_{\mathrm{filter}}^{1/2} = \begin{bmatrix} (S_p\ S_x)^{-1} & \mathrm{Sen}_1\, R_{y_1}^{-1} \\ 0 & R_{y_1}^{-1} \end{bmatrix} \tag{3.34}$$

where

$$P_{(px)} = \begin{bmatrix} P_p & P_{px} \\ P_{xp} & P_x \end{bmatrix} \quad \text{and} \quad P_{(px)y_1} = \begin{bmatrix} P_{py_1} \\ P_{xy_1} \end{bmatrix}$$

To compute the true covariance, we need the covariance of η, $\text{cov}(\eta)$, and this is readily expressed in terms of the previously defined augmented array quantities:

$$\text{cov}(\eta) = \begin{bmatrix} S_{y_2} \\ R_{y_1 y_2} \end{bmatrix} P_{y_2}(0) \begin{bmatrix} S_{y_2} \\ R_{y_1 y_2} \end{bmatrix}^{\mathrm{T}}$$

$$+ \begin{bmatrix} \text{SR}_p & \text{SR}_x & \text{SR}_{y_1} & \text{SR}_{y_2} \\ 0 & 0 & \text{sr}_{y_1} & \text{sr}_{y_1 y_2} \end{bmatrix} \begin{bmatrix} \text{SR}_p & \text{SR}_x & \text{SR}_{y_1} & \text{SR}_{y_2} \\ 0 & 0 & \text{sr}_{y_1} & \text{sr}_{y_1 y_2} \end{bmatrix}^{\mathrm{T}} \tag{3.35}$$

The actual covariance of the filter estimate is (cf. Eq. (3.5))

$$P_{\text{act}} = P_{\text{filt}}^{1/2} \text{cov}(\nu) P_{\text{filt}}^{\mathrm{T}/2} \tag{3.36}$$

It is of interest to note that the form of this result is similar to, but not identical with, the consider formulae (VIII.2.12), (VIII.2.15), (VIII.2.17), and (VIII.2.19).

The important results of this section are the data inclusion and propagation algorithms. They show that the effects of incorrect statistics can be analyzed by simply adjoining additional matrix columns to the square root information array. The form of the algorithm allows one to evaluate bias parameter effects in a unified manner, i.e., the same formulae apply whether or not incorrect *a priori* statistics are involved. It should be noted that in the case that correct *a priori* statistics are used by the filter, $\text{cov}(\eta)$ simplifies to

$$\text{cov}(\eta) = I + \begin{bmatrix} S_{y_2} \\ R_{y_1 y_2} \end{bmatrix} P_{y_2}(0) \begin{bmatrix} S_{y_2} \\ R_{y_1 y_2} \end{bmatrix}^{\mathrm{T}} \tag{3.37}$$

and (3.36) reduces to

$$P_x^{\mathrm{a}} = P_{\text{filt}} + \left\{ P_{\text{filt}}^{1/2} \begin{bmatrix} S_{y_2} \\ R_{y_1 y_2} \end{bmatrix} \right\} P_{y_2}(0) \left\{ P_{\text{filt}}^{1/2} \begin{bmatrix} S_{y_2} \\ R_{y_1 y_2} \end{bmatrix} \right\}^{\mathrm{T}} \tag{3.38}$$

and this result corresponds to that of Eq. (VII.2.18).

Another special case corresponding to (3.36) is the case of only *a priori* errors in the p, x, and y covariances; there are no y_2 parameters. In this case,

$$[P_{\text{act}}^{\text{ap}}]^{1/2} = P_{\text{filt}}^{1/2}\begin{bmatrix} \text{SR}_p & \text{SR}_x & \text{SR}_{y_1} \\ 0 & 0 & \text{sr}_{y_1} \end{bmatrix} \tag{3.39}$$

where the superscript "ap" betokens covariance due to *a priori* errors.

Analyzing the effects of unestimated colored noise is similar to the treatment of unestimated biases. There are, however, some significant differences, i.e.,

(1) Unlike the bias errors, the colored noise covariance perturbations do not have rank one. Thus the effects of a colored noise parameter cannot be analyzed using a sensitivity vector.

(2) The list of colored noise terms (estimated and considered) must be selected at the outset of a simulation. One cannot decide *a posteriori* which colored noise variables will be estimated and which will be considered.

(3) Unestimated colored noise parameters are correlated with the data equation error ν_X. Equations for this correlation must be included.

It turns out that the effects of unestimated colored noise can be analyzed by addending additional matrix columns to the information tableau, and with the exceptions noted above, the derivation is analogous to the bias treatment already given. An equation mechanization is presented in Appendix IX.B, which includes the effects of incorrect covariance *a priori* and unestimated bias parameters that were discussed in Sections IX.2 and IX.3, respectively, as well as the addenda needed to consider the effects of unmodeled colored noise terms.

APPENDIX IX.A Processing of the Augmented Information Array Columns

Our goal is to exploit the form of $\tilde{T}_w$ so that the evaluation of (3.17) can be obtained as an addendum to the filter information array. To accomplish this we exploit the product structure of $\tilde{T}_p$, the diagonal property of SR_w, and the zeros occurring in Eq. (3.17). We begin with

$$\tilde{T}_p\begin{bmatrix} \text{SR}_w & 0 \\ 0 & L_0 \end{bmatrix} = \left[\begin{array}{ccc} * & * & * \\ \hline & F & \end{array}\right]\}N \tag{3.17a}$$

with

$$L_0 = [\widehat{SR}_p \quad \widehat{SR}_x \quad \widehat{SR}_y]; \qquad F = \left[\tilde{L}_w \quad \tilde{L}_p \quad \tilde{L}_x \quad \tilde{L}_y\right]$$
$$SR_w = \text{Diag}(\sigma_w(1), \ldots, \sigma_w(N_p)), \qquad \tilde{T}_p = \tilde{T}_p(N_p) \cdots \tilde{T}_p(1) \tag{A.1}$$

and where $\tilde{T}_p(i)$ is an elementary Householder transformation (cf. Chapter IV)

$$\tilde{T}_p(i) = I - \beta_i u_i u_i^{\mathrm{T}}$$

The β_i are scalars and the u_i have the form

$$u_i = \begin{bmatrix} 0 \\ \lambda(i) \\ 0 \\ v_i \end{bmatrix} \begin{matrix} \}N_p - i \\ \}1 \\ \}i - 1 \\ \}N \end{matrix} \tag{A.2}$$

Because of (A.1) and (3.17a) we can write

$$A_i = \tilde{T}_p(i) A_{i-1}; \quad A_0 = \begin{bmatrix} SR_w & 0 \\ 0 & L_0 \end{bmatrix}$$

and the array A_{N_p} will contain the desired result. The special structure of u_i indicates that at step i only $N + 1$ rows of A_{i-1} are altered. They are the $(N + i)$th row (from the bottom), and the bottom N rows. Further, it is observed that the only rows that participate in the ith step of the computation are the rows that are to be altered. These comments suggest that only $N + 1$ rows are needed for our calculations. The computation is consequently arranged in the following abbreviated form:

$$u_i = \begin{bmatrix} \lambda(i) \\ v_i \end{bmatrix} \begin{matrix} \}1 \\ \}N \end{matrix} \tag{A.3}$$

$$\tilde{T}_i = I_{N+1} - \beta_i \bar{u}_i \bar{u}_i^{\mathrm{T}} \tag{A.4}$$

where $\tilde{T}_i$ is the form of the transformation that is used in the filter array computation of Chapter VI, i.e.,

$$\begin{matrix} 1\{ \\ N\{ \end{matrix}\overbrace{\begin{bmatrix} * - * \\ L_i \end{bmatrix}}^{N+N_y+i} = \tilde{T}_i \begin{bmatrix} \overset{1}{\lambda} & \overbrace{0}^{N+N_y+i-1} \\ 0 & L_{i-1} \end{bmatrix}, \qquad i = 1, \ldots, N_p \tag{A.5}$$

where $N = N_p + N_X$ and $L_{N_p} = F$. Note that a byproduct of result (A.5) is that storage requirements have been reduced.

APPENDIX IX.B The Augmented Array SRIF Error Analysis Algorithm

In this appendix we describe the computer mechanization of the SRIF augmented matrix tableau error analysis. The algorithm to be presented includes the effects of (a) incorrect *a priori* covariances of the initial estimate, measurement errors, and process noise errors, (b) unestimated bias parameters, and (c) unestimated colored noise variables. Consider colored noise terms are distinguished from the estimated colored noise terms in that the former variables carry a subscript "c," i.e.

$$p_c = \begin{bmatrix} p_c(1) \\ \vdots \\ p_c(N_{p_c}) \end{bmatrix} \qquad \text{(unestimated colored noise variables)}$$

$$M_c = \text{Diag}(m_c(j)) = \text{Diag}(\exp(-DT/\tau_c(j)))$$

etc. Our computations involve

$$P_{S_v p_c} = E(S_v p_c^{\mathrm{T}}) = E\left(\begin{bmatrix} \nu_p \\ \nu_X \end{bmatrix} p_c^{\mathrm{T}}\right) \quad \text{and} \quad P_{\nu_y p_c} = E(\nu_y p_c^{\mathrm{T}})$$

Our algorithm implementation utilizes the FORTRAN method of "writing over" old variables. This keeps the notation uncluttered and facilitates computer implementation.

Augmented Information Array

$$[\overbrace{S_p}^{N_p} \quad \overbrace{S_x}^{N_x} \quad \overbrace{S_y}^{N_y} \quad \overbrace{S_z}^{1} \quad \overbrace{S_{p_c}}^{N_{p_c}} \quad \overbrace{\mathrm{SR}_p}^{N_p} \quad \overbrace{\mathrm{SR}_x}^{N_x} \quad \overbrace{\mathrm{SR}_y}^{N_y} \quad \overbrace{P_{S_\nu p_c}}^{N_{p_c}}]\} N_p + N_x \tag{B.1}$$

$$[\overbrace{R_y}^{N_y} \quad \overbrace{z_y}^{1} \quad \overbrace{R_{yp_c}}^{N_{p_c}} \quad \overbrace{\mathrm{sr}_y}^{N_y} \quad \overbrace{P_{\nu_y p_c}}^{N_{p_c}}]\} N_y \tag{B.2}$$

The significance of the tableau elements in (B.1) and (B.2) is

$$\begin{bmatrix} S_p & S_x & S_y & S_{p_c} \\ 0 & 0 & R_y & R_{yp_c} \end{bmatrix} \begin{bmatrix} p \\ x \\ y \\ p_c \end{bmatrix} = \begin{bmatrix} S_z \\ z_y \end{bmatrix} - \begin{bmatrix} S_\nu \\ \nu_y \end{bmatrix}$$

and

$$\left[\operatorname{cov}\begin{pmatrix} S_\nu \\ \nu_y \end{pmatrix}\right]^{1/2} = \begin{bmatrix} \mathrm{SR}_p & \mathrm{SR}_x & \mathrm{SR}_y \\ 0 & 0 & \mathrm{sr}_y \end{bmatrix}$$

$$[S_p \quad S_x \quad S_y \quad S_z] = \begin{bmatrix} R_p & R_{px} & R_{py} & z_p \\ R_{xp} & R_x & R_{xy} & z_x \end{bmatrix} \tag{B.3}$$

Initialization

$$[S_p \quad S_x \quad S_y], \quad R_y \qquad \text{filter initial values} \tag{B.4}$$

$$\begin{bmatrix} \mathrm{SR}_p & \mathrm{SR}_x & \mathrm{SR}_y \\ 0 & 0 & \mathrm{sr}_y \end{bmatrix}$$

$$= \left[\begin{array}{c|c} S_p \; S_x & S_y \\ \hline 0 & R_y \end{array}\right] \left[\begin{array}{c|c} (S_p^{\mathrm{a}} \; S_x^{\mathrm{a}})^{-1} & -(S_p^{\mathrm{a}} \; S_x^{\mathrm{a}})^{-1} S_y^{\mathrm{a}} (R_y^{\mathrm{a}})^{-1} \\ \hline 0 & (R_y^{\mathrm{a}})^{-1} \end{array}\right] \tag{B.5}$$

$$[S_{p_c}\ S_z] = [\mathrm{SR}_p\ \mathrm{SR}_x][S_{p_c}^{\mathrm{a}}\ S_z^{\mathrm{a}}], \qquad [R_{yp_c}\ z_y] = \mathrm{sr}_y[R_{yp_c}^{\mathrm{a}}\ z_y^{\mathrm{a}}] \tag{B.6}$$

$$P_{S_\nu p_c} = -[S_p\ S_x\ S_y]\begin{bmatrix} P_{pp_c}^{\mathrm{a}} \\ P_{xp_c}^{\mathrm{a}} \\ P_{yp_c}^{\mathrm{a}} \end{bmatrix} - S_{p_c}P_c, \quad P_{\nu_y p_c} = -R_y P_{yp_c}^{\mathrm{a}} - R_{yp_c}P_c \tag{B.7}$$

Square root covariance of the process noise data equation error:

$$\mathrm{SR}_w = \mathrm{Diag}\left[R_w(1)(R_w^{\mathrm{a}}(1))^{-1}, \ldots, R_w(N_p)\left(R_w^{\mathrm{a}}(N_p)\right)^{-1}\right] \tag{B.8}$$

Square root covariance of the measurement equation error:

$$\mathrm{SR}_\nu = R_\nu (R_\nu^{\mathrm{a}})^{-1} \tag{B.9}$$

Propagation

$$\begin{aligned} x_{j+1} &= V_x x_j + V_{p_c} p_{c\,|\,t_j} + V_p p_j + V_y y \\ p_{j+1} &= M p_j + w_j, \qquad p_{c\,|\,t_{j+1}} = M_c p_{c\,|\,t_j} + w_{c\,|\,t_j} \end{aligned} \tag{B.10}$$

The mapping (B.10) gives rise to the following information array modifications:

$$\begin{aligned} S_x &:= S_x V_x^{-1}, \qquad S_y := S_y - S_x V_y \\ L_0 &= [\mathrm{SR}_p \quad \mathrm{SR}_x \quad \mathrm{SR}_y \quad P_{S_\nu p_c}] \end{aligned} \tag{B.11}$$

For $k = 1, \ldots, N_p$, evaluate Eqs. (B.12a) and (B.12b):

$$S_p(1) = S_p(1) - S_x V_p(k) \tag{B.12a}$$

$$\tilde{T}_p(k)\left[\begin{array}{cc|cccccc|cc} \overbrace{-R_w(k)m(k)}^{1} & \overbrace{0}^{N_p-1} & R_w(k) & 0 & 0 & z_w(k) & 0 & & \overbrace{\mathrm{SR}_w(k) \quad 0}^{N+N_{p_c}+k-1} & \\ \multicolumn{2}{c|}{S_p} & 0 & S_x & S_y & S_z & S_{p_c} & & 0 & L_{k-1} \end{array}\right]$$

$$:= \left[\begin{array}{cc|cccc|c} * & * & * & * & * & * & \overbrace{* - *}^{N+N_{p_c}+k} \\ 0 & S_p & S_x & S_y & S_z & S_{p_c} & L_k \end{array}\right] \tag{B.12b}$$

where $N = N_p + N_x + N_y$ and $\tilde{T}_p(k)$ is an elementary Householder transformation.

$$L_{N_p} = [L_w \quad L_p \quad L_x \quad \mathrm{SR}_y \quad P_{S_v p_c}] \tag{B.13}$$

$$S_{p_c} := (S_{p_c} - S_x V_{p_c}) M_c^{-1}, \qquad R_{yp_c} := R_{yp_c} M_c^{-1} \tag{B.14}$$

(M_c is, in most cases, diagonal)

$$\begin{bmatrix} P_{S_v p_c} \\ P_{v_y p_c} \end{bmatrix} := \begin{bmatrix} P_{S_v p_c} \\ P_{v_y p_c} \end{bmatrix} M_c^{\mathrm{T}} - \begin{bmatrix} S_{p_c} \\ R_{yp_c} \end{bmatrix} P_{w_c} \tag{B.15}$$

$$\left[\begin{array}{c|ccc} & \mathrm{SR}_p & \mathrm{SR}_x & \mathrm{SR}_y \\ 0 & & & \\ & 0 & 0 & \mathrm{sr}_y \end{array}\right]$$

$$:= \begin{bmatrix} L_w & L_p & L_x & \mathrm{SR}_y & S_{p_c}(\mathrm{SR}_{w_c}) \\ 0 & 0 & 0 & \mathrm{sr}_y & R_{yp_c}(\mathrm{SR}_{w_c}) \end{bmatrix} \tilde{T}_{pxy} \tag{B.16}$$

where $\tilde{T}_{pxy}$ is orthogonal and SR_p is upper triangular.

Measurement Processing

$$z = A_p p + A_x x + A_y y + A_{p_c} p_c + v \tag{B.17}$$

$$\tilde{T}_{px} \begin{bmatrix} S_p & S_x & S_y & S_z & S_{p_c} & \mathrm{SR}_p & \mathrm{SR}_x & \mathrm{SR}_y & P_{S_v p_c} & 0 \\ A_p & A_x & A_y & z & A_{p_c} & 0 & 0 & 0 & 0 & \mathrm{SR}_v \end{bmatrix}$$

$$:= \begin{bmatrix} S_p & S_x & S_y & S_z & S_{p_c} & L_p & L_x & L_y & P_{S_v p_c} & L_v \\ 0 & 0 & \hat{A}_y & \hat{z} & \hat{A}_{p_c} & M_p & M_x & M_y & M_{p_c} & M_v \end{bmatrix} \tag{B.18}$$

$$\hat{T}_y \begin{bmatrix} R_y & z_y & R_{yp_c} & 0 & 0 & \text{sr}_y & 0 & P_{\nu_y p_c} \\ \hat{A}_z & \hat{z} & \hat{A}_{p_c} & M_p & M_x & M_y & M_\nu & M_{p_c} \end{bmatrix}$$

$$:= \begin{bmatrix} R_y & z_y & R_{yp_c} & H_p & H_x & H_y & H_\nu & P_{\nu_y p_c} \\ 0 & * & * & * & * & * & * & * \end{bmatrix} \qquad \text{(B.19)}$$

$$\left[\begin{array}{c|ccc} \overbrace{0}^{m} & \text{SR}_p & \text{SR}_x & \text{SR}_y \\ & 0 & 0 & \text{sr}_y \end{array}\right] := \begin{bmatrix} L_p & L_x & L_y & L_\nu \\ H_p & H_x & H_y & H_\nu \end{bmatrix} T_\nu \qquad \text{(B.20)}$$

When desired, estimates and covariances can be obtained as follows. Let $y = \left[\begin{smallmatrix} y_1 \\ y_2 \end{smallmatrix}\right]$, y_1 estimated and y_2 considered:

$$[\text{Sen}_1\ \text{Sen}_2] = -[S_p\ S_x]^{-1}[S_{y_1}\ S_{y_2}]$$

$$R_y = \begin{bmatrix} R_{y_1} & R_{y_1 y_2} \\ 0 & R_{y_2} \end{bmatrix}, \qquad z_y = \begin{bmatrix} z_{y_1} \\ z_{y_2} \end{bmatrix} \qquad \text{(B.21)}$$

$$\hat{y}_1 = R_{y_1}^{-1} z_{y_1}, \qquad \begin{bmatrix} p \\ x \end{bmatrix}_{\text{est}} = [S_p\ S_x]^{-1} S_z + \text{Sen}_1 \hat{y}_1 \qquad \text{(B.22)}$$

$$P_{\text{filt}}^{1/2} = \begin{bmatrix} [S_p\ S_x]^{-1} & \text{Sen}_1 R_{y_1}^{-1} \\ 0 & R_{y_1}^{-1} \end{bmatrix}, \qquad P_{\text{filt}} = P_{\text{filt}}^{1/2} P_{\text{filt}}^{\text{T}/2} \qquad \text{(B.23)}$$

P_{filt} is the computed covariance of

$$\begin{bmatrix} p \\ x \\ y_1 \end{bmatrix}_{\text{est}}$$

For the actual covariance of

$$\begin{bmatrix} p \\ x \\ y_1 \end{bmatrix}$$

use

$$P_{\text{act}} = P_{\text{filt}}^{1/2} \operatorname{cov} \eta \, P_{\text{filt}}^{\text{T}/2} \tag{B.24}$$

where

$$\begin{aligned} \operatorname{cov}(\eta) = {} & \begin{bmatrix} S_{y_2} \\ R_{y_1 y_2} \end{bmatrix} P_{y_2}(0) \begin{bmatrix} S_{y_2} \\ R_{y_1 y_2} \end{bmatrix}^{\text{T}} + \begin{bmatrix} S_{p_c} \\ R_{y p_c} \end{bmatrix} P_c \begin{bmatrix} S_{p_c} \\ R_{y p_c} \end{bmatrix}^{\text{T}} \\ & + \begin{bmatrix} \mathrm{SR}_p & \mathrm{SR}_x & \mathrm{SR}_{y_1} & \mathrm{SR}_{y_2} \\ 0 & 0 & \mathrm{sr}_{y_1} & \mathrm{sr}_{y_1 y_2} \end{bmatrix} \begin{bmatrix} \mathrm{SR}_p & \mathrm{SR}_x & \mathrm{SR}_{y_1} & \mathrm{SR}_{y_2} \\ 0 & 0 & \mathrm{sr}_{y_1} & \mathrm{sr}_{y_1 y_2} \end{bmatrix}^{\text{T}} \\ & + \begin{bmatrix} P_{s_v p_c} \\ P_{v_{y_1} p_c} \end{bmatrix} \begin{bmatrix} S_{p_c} \\ R_{y_1 p_c} \end{bmatrix}^{\text{T}} + \left\{ \begin{bmatrix} P_{s_v p_c} \\ P_{v_{y_1} p_c} \end{bmatrix} \begin{bmatrix} S_{p_c} \\ R_{y_1 p_c} \end{bmatrix}^{\text{T}} \right\}^{\text{T}} \end{aligned} \tag{B.25}$$

Note that the mechanization simplifies considerably when the colored noise terms are omitted from the consider analysis. If only the effects of incorrect covariance *a priori* and bias consider parameters are included in the mechanization, one can approximate the effects of unestimated colored noise. To do this, use a filter with small *a priori* standard deviations for the colored noise parameters that are to be considered. It is generally sufficient to use a filter sigma that is negligible compared with the actual value; viz.: take $\sigma(\text{filter}) < \epsilon\ \sigma(\text{actual})$, with ϵ as the machine tolerance. It is our experience that consider colored noise variables can be quite adequately analyzed using this method and that there is no need to encumber a computer implementation with inessentials. The results of Sections IX.2 and IX.3 were extended here to include consider colored noise variables mostly for pedagogical reasons; however, we believe that some accuracy analyses might benefit from this mechanization.

References

[1] Nishimura, T., and Nead, M., Athena filter sequential orbit determination program with general evaluation capability, JPL Tech. Rep. No. 900–605 (March 1973).

A general algorithm is presented for covariance evaluation of a SRIF that uses incorrect *a priori*. The evaluation algorithm includes considering the effects of unmodeled parameters. The analysis consists, in the main, of manipulating Kalman covariance error analysis results (many of which were first derived by Nishimura) and arranging the evaluation algorithm to exploit the numerical attributes of the Householder transformation. The mathematical development is brief, with most of the emphasis being put on the computer program user description.

[2] Griffin, R. E., and Sage, A. P., Sensitivity analysis of discrete filtering and smoothing algorithms, *AIAA J.* **7**, No. 10 1890–1897 (1969).

Error analysis results of a very general nature are derived both for filtering and for smoothing. Their algorithms are, however, presented without special regard for computational efficiency or storage economy.

[3] Curkendall, D. W., Problems in estimation theory with applications to orbit determination, UCLA School Eng. and Appl. Sci. Rep., No. UCLA-ENG-7275 (September 1972).

This thesis introduces SRIF error analysis. The results and method described here strongly influenced our development of Section IX.3. Our technique (cf. Section 3) was in fact motivated by Curkendall's approach. His vivid exposition is easy to grasp; however, his proposed implementation formulae are lengthier and less efficient than are the comparable results presented here.

X

Square Root Information Smoothing

X.1 Introduction

Our estimation development up to this point has been directed toward the filtering and prediction problems; i.e., data up until the current time was used to estimate current and future states of our system model. Although our stress has been on the filtering problem we have, in Chapter VI, solved the prediction problem. The results presented there show, as one might expect, that prediction corresponds simply to projecting the latest filter estimate ahead in time. Having dealt with the problems of filtering and prediction we turn our energies, in this chapter, to the solution of the smoothing problem; i.e., the use of past and current data to "best" estimate the past history of our system.

One simple but effective method of attacking the smoothing problem is to embed it into a filtering problem. For example, the model $x_{j+1}=\Phi_j x_j + G_j w_j$ can be embedded into an augmented system of the same form (cf. Willman (1969)):

$$X_{j+1}^{\text{Tot}} = \Phi_j^{\text{Tot}} X_j^{\text{Tot}} + G_j^{\text{Tot}} w_j \tag{1.1}$$

where

$$X_j^{\text{Tot}} = \begin{bmatrix} x_0 \\ \vdots \\ x_N \end{bmatrix}; \qquad G_j^{\text{Tot}} = \begin{bmatrix} 0 \\ \vdots \\ 0 \\ G_j \\ 0 \\ \vdots \\ 0 \end{bmatrix} \begin{matrix} \left.\vphantom{\begin{matrix}0\\ \vdots \\ 0\end{matrix}}\right\} j-1 \\ \\ \\ \\ \\ \end{matrix}$$

and

$$\Phi_j^{\text{Tot}} = \left[\begin{array}{c|c|c} \overbrace{\begin{matrix} I & & \\ & \ddots & \\ & & I\end{matrix}}^{j-1} & \overbrace{\begin{matrix}0\\ \vdots \\ 0\end{matrix}}^{l^\dagger} & 0 \\ \hline 0 \cdots \Phi_j & 0 & 0\cdots 0 \\ \hline 0 & \begin{matrix}0\\ \vdots\\ 0\end{matrix} & \begin{matrix} I & & \\ & \ddots & \\ & & I\end{matrix}\end{array}\right] \begin{matrix} \}\; j-1 \\ \}\; l^\dagger \\ \\ \end{matrix}$$

The key to understanding how filtering this system solves the smoothing problem lies in the observation that for constant parameters the concepts of filtering and smoothing coincide. Since the first k (vector) components of X_j^{Tot} are constant for $j \geqslant k$ it follows readily that the filtered estimates of these components will also be smoothed estimates. To derive covariance smoothing algorithms one could introduce matrix partitioning into the Kalman filter equations corresponding to this model. Algorithms so derived create smoothed estimates as a function of the incoming data. Such algorithms, while useful and interesting, require excessive amounts of computer storage and computation.‡ We shall be interested in a slightly different problem formulation, one that requires essentially no additional computer storage and only a modicum of additional computation; i.e., the determination of smoothed state vector histories related to a given arc of measurement data. In the literature this problem is called "fixed interval" smoothing. Our approach to fixed interval smoothing is to consider it as part of the least squares performance minimization:

$$J^{(N+1)} = \|\tilde{R}_0 x_0 - \tilde{z}_0\|^2 + \sum_{j=0}^{N} \left(\|A_j x_j - z_j\|^2 + \|R_w(j) w_j - z_w(j)\|^2\right) \tag{VI.2.24}$$

†Note that the elements of Φ_j^{Tot} are matrices, so that the "l" represents a thickness having length the dimension of x.

‡The significance of (1.1), for us, is that it enables us to consider smoothing as a filtering problem; as a result, filter concepts, such as sensitivity and consider parameter analyses, can be transferred to the smoothing problem.

For our purposes, smoothed estimates $x_j^*, j = 0, \ldots, N$, can be defined as the values that minimize $J^{(N+1)}$ subject to the dynamics constraint

$$x_{j+1} = \Phi_j x_j + G_j w_j \qquad \text{(VI.1.1)}$$

One of the benefits accruing to identifying smoothing as a filtering problem is that we can draw on our (linear) filtering experience and conclude that smoothed estimates as just defined are also minimum variance, i.e., x_j^* minimizes

$$\min_x E\left[(x_j - x)^{\mathrm{T}}(x_j - x)\middle|z_k : k = 0, \ldots, N\right]$$

$$= \min_x \operatorname{Tr} E\left[(x_j - x)(x_j - x)^{\mathrm{T}}|z_k : k = 0, \ldots, N\right]$$

and this means that $x_j^* = x_{j/N}$ gives a minimum *a posteriori* covariance.

Space navigation analysts at JPL and elsewhere have traditionally formulated their problems in terms of least squares criteria, but have generally chosen to refer their variables to the epoch state of the system. Until recently process noise was omitted from such problems. The filter technology popularized by Kalman allowed analysts to better model their physical problems, and space missions which involved stringent navigation accuracy requirements necessitated use of these models. Because the Kalman filter is generally used to obtain filtered estimates of the current model state, it was inferred that the epoch state formulation would have to be abandoned. There was, however, a reluctance on the part of space navigation analysts to reformulate their problems in terms of the current state.† The pseudoepoch state, Eq. (VII.B.2), solved this apparent dilemma because it allowed the inclusion of process noise and at the same time preserved the epoch state formulation

$$x(t) = \text{current state} = V_x(t_j, t_0)x_j + V_y(t_j, t_0)y \qquad \text{(VII.B.2)}$$

This model formulation is compatible with previously developed software, and in terms of our applications, it leads to a computationally efficient filter mechanization. It was gratifying to discover (see Appendix X.B) that use of this model reduces smoothing computations almost to the point of triviality. A summary of the topics to be discussed in this chapter follows.

†Reasons for such reluctance include inertia (human nature tends to resist change) and cost (much of the previously developed operational software would require revision, and this would involve lengthy expenditures of time and manpower).

Various recursions which solve the smoothing problem are presented in Section X.2. Estimates and estimate error covariances are obtained and these are used in Appendix X.A to derive the classic results of Rauch, Tung, and Streibel (RTS) [1]. This appendix also contains a discussion of the modified Bryson–Frazier (mBF) adjoint variable smoothing algorithm [2]. It is much more efficient and amenable to computer implementation than is the RTS result. In Section X.3 we pay special attention to systems composed partly of biases because most of our applications are of this type. A unified treatment of biases is demonstrated, i.e., it is shown that a single set of formulae can be used to express filter and smoother sensitivity and consider analysis results. Appendix X.B contains an elegant quasi-FORTRAN mechanization of the Dyer–McReynolds covariance smoother† (DMCS), which ties together the epoch state model used for the SRIF mechanization, Appendix VII.B, and the treatment of biases that is discussed in Section X.3.

It is generally accepted that the SRIF–DMCS approach to estimation is more stable, numerically accurate, and flexible in its treatment of systems composed partially of biases than are the other methods currently enjoying great popularity in estimation circles. The SRIF computer implementation (Appendix VII.B), the augmented array approach to SRIF error analysis (Chapter IX), and the DMCS mechanization (Appendix X.B) demonstrate that for a wide class of problems this improved stability, accuracy, and flexibility can be achieved without sacrificing computational efficiency. For the class of problems encountered in space navigation our experience indicates that the SRIF–DMCS computer implementation is competitive with the Kalman covariance implementation (in terms of computation and computer storage requirements).

X.2 Square Root and Non-Square Root Smoother Recursions

As mentioned in the introduction, our smoother analysis is predicated on minimizing the general least squares performance functional $J^{(N+1)}$, Eq. (VI.2.25). It was a simple matter to express $J^{(N+1)}$ in terms of the SRIF tableau entries, i.e.,

$$J^{(N+1)} = \sum_{j=0}^{N} \|e_j\|^2 + \sum_{j=0}^{N} \|\tilde{R}_w(j)w_j + \tilde{R}_{wx}(j)x_{j+1} - \tilde{z}_w(j)\|^2 + \|\hat{R}_{N+1}x_{N+1} - \tilde{z}_{N+1}\|^2 \qquad \text{(VI.2.27)}$$

†The names Dyer and McReynolds are attached to the estimate and covariance recursions, (2.2) and (2.10), not because either of them fabricated these relations, but because the algorithms depend on quantities that are computed via the Dyer–McReynolds square root filtering algorithm.

Since the e's are independent of the w and x sequence elements, it is clear that $J^{(N+1)}$ is minimized if the norm terms involving the x's are made to vanish. When this condition is combined with the model constraint (VI.1.1), the smoothed state recursion results:

$$x_{N+1}^* = \tilde{R}_{N+1}^{-1}\tilde{z}_{N+1} \tag{2.1}$$

$$\left.\begin{aligned} w_j^* &= \left[\tilde{R}_w(j)\right]^{-1}\left(\tilde{z}_w(j) - \tilde{R}_{wx}(j)x_{j+1}^*\right) \\ x_j^* &= \Phi_j^{-1}(x_{j+1}^* - Gw_j^*) \end{aligned}\right\} j = N, N-1, \ldots, 0 \tag{2.2}$$

Remark: Note that $\tilde{R}_w$ is either a triangular matrix or a scalar (when process noise is included one component at a time). In either case computation is facilitated.

To obtain covariances for the smoothed estimates we employ reasoning that parallels our development of the SRIF process noise algorithm. The key idea, central to our earlier developments, is to represent the estimate and covariance in terms of an equivalent augmented state data equation. The pattern becomes transparent once the first step is outlined; so without further ado we proceed to describe this first step (at time N) and write the data equations for w_N and x_{N+1}:

$$\begin{aligned} \tilde{R}_w(N)w_N + \tilde{R}_{wx}(N)x_{N+1} &= \tilde{z}_w(N) - \tilde{\nu}_w(N) \\ R_x^*(N+1)x_{N+1} &= z_x^*(N+1) - \nu_x^*(N+1) \end{aligned} \tag{2.3}$$

where $\tilde{\nu}_w \in N(0, I)$ and $\tilde{R}_x(N+1)$ and $z_x(N+1)$ have been replaced by asterisk notation (smoothed and filtered values coincide at the terminal time). Replacing x_{N+1} by the mapping equation (VI.1.1), we obtain

$$\begin{bmatrix} \tilde{R}_w(N) + \tilde{R}_{wx}(N)G & \tilde{R}_{wx}(N)\Phi_N \\ R_x^*(N+1)G & R_x^*(N+1)\Phi_N \end{bmatrix}\begin{bmatrix} w_N \\ x_N \end{bmatrix}$$

$$= \begin{bmatrix} \tilde{z}_w(N) \\ z_x(N+1) \end{bmatrix} - \begin{bmatrix} \tilde{\nu}_w(N) \\ \nu_x^*(N+1) \end{bmatrix} \tag{2.4}$$

Since the ν terms have unity covariances we are motivated to apply our previously successful method (cf. Chapter VI) of using orthogonal transformations to eliminate the w variables. Thus we choose an orthogonal transformation T_N^* to partially triangularize (2.4), and represent the result in data array form.

$$T_N^* \begin{bmatrix} \tilde{R}_w(N) + \tilde{R}_{wx}(N)G & \tilde{R}_w(N)\Phi_N & \tilde{z}_w(N) \\ R_x^*(N+1)G & R_x^*(N+1)\Phi_N & z_x^*(N+1) \end{bmatrix}$$

$$= \begin{bmatrix} R_w^*(N) & R_{wx}^*(N) & z_w^*(N) \\ 0 & R_x^*(N) & z_x^*(N) \end{bmatrix} \tag{2.5}$$

The bottom row of the right side of (2.5) represents the data equation for the smoothed estimate of x_N,

$$R_x^*(N)x_N = z_x^*(N) - \nu_x^*(N) \tag{2.6}$$

Comparing this equation with (2.3) it is easy to see that the process can be repeated, i.e., use (2.6) and the data equation for w_{N-1} to obtain a data equation for x_{N-1}; etc. Thus we have derived the square root information smoother algorithm (SRIS)†:

$$T_j^* \begin{bmatrix} \tilde{R}_w(j) + \tilde{R}_{wx}(j)G & \tilde{R}_{wx}(j)\Phi_j & \tilde{z}_w(j) \\ R_x^*(j+1)G & R_x^*(j+1)\Phi_j & z_x^*(j+1) \end{bmatrix}$$

$$= \begin{bmatrix} R_w^*(j) & R_{wx}^*(j) & z_w^*(j) \\ 0 & R_x^*(j) & z_x^*(j) \end{bmatrix}, \qquad j = N, \ldots, 0 \tag{2.7}$$

At times where estimates and/or covariances are desired we have

$$x_j^* = [R_x^*(j)]^{-1} z_x^*(j); \qquad P_x^*(j) = [R_x^*(j)]^{-1}[R_x^*(j)]^{-\mathrm{T}} \tag{2.8}$$

The SRIS recursion (2.7) requires considerably more computation than does the simple estimate recursion (2.2). This observation motivates us to investigate a more direct means of computing smoothed covariances. A modest manipulation of the w data equation and the mapping gives

$$x_j = \Phi_j^{-1}\left[\left(I + L_j\tilde{R}_{wx}(j)\right)x_{j+1} - L_j\tilde{z}_w(j)\right] - \Phi_j^{-1}\left[\tilde{R}_w(j)\right]^{-1}\tilde{\nu}_w(j) \tag{2.9}$$

where $L_j = G[\tilde{R}_w(j)]^{-1}$. From this we obtain the Dyer–McReynolds covariance smoother (DMCS) recursion

$$P_x^*(j) = \Phi_j^{-1}\left[\left(I + L_j\tilde{R}_{wx}(j)\right)P_x^*(j+1)\left(I + L_jR_{wx}(j)\right)^{\mathrm{T}} + L_jL_j^{\mathrm{T}}\right]\Phi_j^{-\mathrm{T}} \tag{2.10}$$

This smoothed covariance recursion is due to Bierman [5]; and despite its

†This result is attributed to Kaminski (cf. Kaminski and Bryson (1972)).

perhaps ungainly appearance it is more efficient computationally than is the SRIS of Eq. (2.7).

Remark: As a matter of expositional convenience we will generally include the estimate recursion (2.2) in our references to the DMCS.

When the epoch state model with the inclusion of colored process noise one component at a time are used for the SRIF mechanization, the smoother covariance recursion (2.10) computation reduces to a near triviality (cf. Appendix X.B).

X.3 The Treatment of Biases in Square Root Smoothing

In this section the smoothing results are arranged to separate out the effects of the bias parameters. Our aim is to derive smoothing results that parallel the filter results developed in Chapters VII and IX. In particular, we will derive a smoothed "sensitivity" Sen*, and a smoothed "computed" covariance P^{c*}. Advantages of our smoothing results include computational efficiency and a unified approach to filtering and smoothing.

We suppose for simplicity of exposition that the state vector $\begin{bmatrix} X \\ y \end{bmatrix}$ has dynamics of the form

$$\begin{bmatrix} X \\ y \end{bmatrix}_{j+1} = \begin{bmatrix} \Phi_X(j) & \Phi_{Xy}(j) \\ 0 & I \end{bmatrix} \begin{bmatrix} X \\ y \end{bmatrix}_j + \begin{bmatrix} G_j w_j \\ 0 \end{bmatrix} \tag{3.1}$$

where y is a vector of bias parameters. The data equation for $\begin{bmatrix} X \\ y \end{bmatrix}$ that corresponds to the top row of (VI.2.29) is

$$\begin{aligned} &\tilde{R}_w(j+1)w_j + \tilde{R}_{wX}(j+1)X_{j+1} + \tilde{R}_{wy}(j+1)y \\ &\quad = \tilde{z}_w(j+1) - \tilde{\nu}_w(j+1), \qquad \tilde{\nu}_w \in N(0, I) \end{aligned} \tag{3.2}$$

Our intention is to derive the smoothed estimate as a sum of two independent terms, one that depends (linearly) on y and one that does not. With this in mind, we set

$$X_j = X_j^c + X_j^y \tag{3.3}$$

and decompose (3.1) and (3.2) as follows:

$$\tilde{R}_w(j+1)w_j^c + \tilde{R}_{wX}(j+1)X_{j+1}^c = \tilde{z}_w(j+1) - \tilde{\nu}_w(j+1) \tag{3.4}$$

$$X_{j+1}^c = \Phi_X(j)X_j^c + Gw_j^c \tag{3.5}$$

$$\tilde{R}_w(j+1)w_j^y + \tilde{R}_{wX}(j+1)X_{j+1}^y + \tilde{R}_{wy}(j+1)y = 0 \tag{3.6}$$

$$X_{j+1}^y = \Phi_X(j)X_j^y + \Phi_{Xy}(j)y + Gw_j^y \tag{3.7}$$

where $w_j = w_j^c + w_j^y$.

Remark: The approach here is patterned after that used to treat the filtering problem in Section VIII.3.

Equations (3.4) and (3.5) have precisely the same form as that discussed in Section X.2, so that the DMCS results (2.2) and (2.10) (or the SRIS (2.7)) may be applied. Thus the computed smoothed estimates and covariances correspond to solving a smoothing problem with no biases present. The equations involving X^y and w^y, (3.6) and (3.7), are solved in a somewhat different fashion. Since by construction X^y is a linear function of y we write

$$X_j^y = \mathrm{Sen}_j^* y \tag{3.8}$$

and substitute this into (3.6) and (3.7). The resultant recursion for Sen* is written as

$$\mathrm{Sen}_j^* = C_j\, \mathrm{Sen}_{j+1}^* + [\Phi_X(j)]^{-1}\left[L_j \tilde{R}_{wy}(j+1) - \Phi_{Xy}(j)\right] \tag{3.9}$$

where C_j is the Rauch *et al.* smoother gain (cf. Eqs. (X.A.5) and (X.A.6))

$$C_j = [\Phi_X(j)]^{-1}\left(I + L_j \tilde{R}_{wX}(j+1)\right) = \hat{P}_j^c [\Phi_X(j)]^{\mathrm{T}}\left[\tilde{P}_{j+1}^c\right]^{-1} \tag{3.10}$$

where $L_j = G\tilde{R}_w^{-1}(j+1)$ is the same expression used in Eqs. (2.9) and (2.10).

Let us review and examine how the "c" and "y" subscript recursions are used for smoothing.

(a) *Initialization* (at the terminal time N): Consistent with the decomposition (3.3), we set

$$\mathrm{Sen}_N^* = -[R_X^*(N)]^{-1} R_{Xy}^*(N), \quad X_N^{c^*} = X_N^c \quad \left(= [R_X^*(N)]^{-1} z_X^*(N)\right) \tag{3.11}$$

i.e., Sen_N^* and $X_N^{c^*}$ are the filter sensitivity and computed estimate. Corresponding to the initialization of $X_N^{c^*}$ we have

$$P_N^{c^*} = R_X^*(N)^{-1}[R_X^*(N)]^{-\mathrm{T}} \tag{3.12}$$

(b) *Smoothed estimates*: From (3.3) and (3.8) we obtain

$$X_j^* = X_j^{c^*} + \mathrm{Sen}_j^* \hat{y} \tag{3.13}$$

where X^{c^*} can be recursively generated from (2.2), and $\hat{y}$ is the filtered (and also smoothed) estimate of the y bias parameters, based on all of the measurements.

(c) *Smoothed covariances*: From (2.10) and (3.8) we obtain

$$P_X^*(j) = P_j^{c^*} + (\mathrm{Sen}_j^*)\hat{P}_y(\mathrm{Sen}_j^*)^{\mathrm{T}} \tag{3.14}$$

where P^{c^*} can be recursively computed using (2.10) and $\tilde{P}_y$ is the y filter estimate error covariance $[R_y(N)]^{-1}[R_y(N)]^{-T}$.

(d) *Smoothed consider covariance*: When the y parameters are not modeled in the filter the true covariance of the estimate is given by

$$P^*_{\text{CON}}(j) = E\Big[\big(X_j - X_j^{c^*}\big)\big(X_j - X_j^{c^*}\big)^{\mathrm{T}}\Big] = P_j^{c^*} + [\text{Sen}_j^*]P_y(0)[\text{Sen}_j^*]^{\mathrm{T}} \tag{3.15}$$

Observe that these results, (3.12)–(3.15), are of precisely the same form as obtained for filtering a system composed partly of biases (refer to Section VII.2). The generalization described in Section IX.3, solving for some of the y parameters and considering the rest, can be carried over to the smoothing solution *mutatis mutandis*. Indeed, when the results are described in terms of "computed" terms and "sensitivities," it follows that a single set of formulae can be used to obtain estimates and covariances for both the filter and the smoother.

One can also obtain P^{c^*} and Sen* by partitioning the SRIS algorithm (2.7); the result is presented as Eqs. (3.16)–(3.19):

$$T_X^*(\bar{j})\begin{bmatrix} \overbrace{\tilde{R}_w(\bar{j}) + \tilde{R}_{wX}(\bar{j})G}^{N_w} & \overbrace{\tilde{R}_{wX}(\bar{j})\Phi_X(j)}^{N_X} & \overbrace{\tilde{R}_{wX}(\bar{j})\Phi_{Xy}(j) + \tilde{R}_{wy}(\bar{j})}^{N_y} & \overbrace{\tilde{z}_x(\bar{j})}^{1} \\ R_X^*(\bar{j})G & R_X^*(\bar{j})\Phi_X(j) & R_X^*(\bar{j})\Phi_{Xy}(j) + R_{Xy}^*(\bar{j}) & z_X^*(\bar{j}) \end{bmatrix}$$

$$= \begin{bmatrix} R_w^*(\bar{j}) & R_{wX}^*(\bar{j}) & R_{wy}^*(\bar{j}) & z_w^*(\bar{j}) \\ 0 & R_X^*(j) & R_{Xy}^*(j) & z_X^*(j) \end{bmatrix} \tag{3.16}$$

where $\bar{j} = j + 1$ and $T_X^*(\bar{j})$ is an orthogonal transformation chosen to partially triangularize the array. The entries of the bottom row correspond to the smoothed data equation, and these may be used directly in the formulae of Section IX.3; in particular,

$$\text{Sen}_j^* = -[R_X^*(j)]^{-1}R_{Xy}^*(j) \tag{3.17}$$

$$X_j^{c^*} = [R_X^*(j)]^{-1}z_X^*(j) \tag{3.18}$$

$$P_j^{c^*} = [R_X^*(j)]^{-1}[R_X^*(j)]^{-\mathrm{T}} \tag{3.19}$$

The sensitivity and P^{c^*} recursions require less computation than does the partitioned SRIS recursion (3.16). The computer cost differential is, however, generally not significant unless estimates and/or covariances are requested frequently, say at each step, or the dimension of the problem is large. On the other hand, we are reluctant to endorse the SRIS formulation because it does not appear as if the X^{c^*}, P^{c^*}, and Sen* recursions are prone to numerical instability.

Appendix X.A Kalman Filter Related Smoothers

The SRIS, Eq. (X.2.7) and the DMCS recursions, (X.2.2) and (X.2.10), depend explicitly on entries of the SRIF tableau. In this appendix we solve the smoothing problem for those cases where a covariance filter is used. To apply the DMCS results in such cases it is necessary to eliminate the SRIF entries from the algorithm, i.e., $L_j = G[\tilde{R}_w(j)]^{-1}$ and $I + L_j\tilde{R}_{wx}(j)$ must be expressed in terms of filter covariances. Our approach is simply to partition and expand the SRIF propagation algorithm (VI.2.29). The technique parallels the method used in Section VI.3 to show the equivalence of the SRIF and Kalman filter propagation algorithms. Thus we are to expand

$$\tilde{T}\begin{bmatrix} R_w & 0 \\ -\hat{R}\Phi^{-1}G & \hat{R}\Phi^{-1} \end{bmatrix} = \begin{bmatrix} \tilde{R}_w & \tilde{R}_{wx} \\ 0 & \tilde{R} \end{bmatrix} \tag{VI.2.29}$$

where to simplify the notation we write $\hat{R} = \hat{R}_x(j)$, $\Phi = \Phi_j$, $\tilde{R}_w = \tilde{R}_w(j+1)$, $\tilde{R}_{wx} = \tilde{R}_{wx}(j+1)$, and $\tilde{R} = \tilde{R}_x(j+1)$.

Because $\tilde{T}$ is orthogonal ($\tilde{T}^{\mathrm{T}}\tilde{T} = I$), we can also write

$$\begin{bmatrix} R_w & 0 \\ -\hat{R}\Phi^{-1}G & \hat{R}\Phi^{-1} \end{bmatrix} = \tilde{T}^{\mathrm{T}}\begin{bmatrix} \tilde{R}_w & \tilde{R}_{wx} \\ 0 & \tilde{R} \end{bmatrix} \tag{A.1}$$

Suppose that $\tilde{T}$ is partitioned consistent with (A.1),

$$\tilde{T} = \begin{matrix} \overbrace{\phantom{T_{11}}}^{N_w} & \overbrace{\phantom{T_{12}}}^{N_x} & \\ \left[\begin{matrix} \tilde{T}_{11} \\ \tilde{T}_{21} \end{matrix}\right. & \left.\begin{matrix} \tilde{T}_{12} \\ \tilde{T}_{22} \end{matrix}\right] & \begin{matrix} \}N_w \\ \}N_x \end{matrix} \end{matrix}$$

Then, from (A.1) we obtain

$$-\hat{R}\Phi^{-1}G = \tilde{T}_{12}^{\mathrm{T}}\tilde{R}_w$$

Thus

$$\tilde{T}_{12}^{\mathrm{T}} = -\hat{R}\Phi^{-1}G\tilde{R}_w^{-1} \tag{A.2}$$

The following group of equations is self-explanatory:

$$\tilde{T}_{12}^{\mathrm{T}}\tilde{R}_{wx} = -\hat{R}\Phi^{-1}G\tilde{R}_w^{-1}\tilde{R}_{wx} \quad \left(\text{multiply (A.2) by } \tilde{R}_{wx}\right)$$

$$\tilde{T}_{12}^{\mathrm{T}}\left(\tilde{T}_{12}\hat{R}\Phi^{-1}\right) = -\hat{R}\Phi^{-1}G\tilde{R}_w^{-1}\tilde{R}_{wx} \quad \left(\text{cf. } \tilde{R}_{wx} \text{ in Eq. (VI.2.29)}\right)$$

$$\left(I - \tilde{T}_{22}^{\mathrm{T}}\tilde{T}_{22}\right)\hat{R}\Phi^{-1} = -\hat{R}\Phi^{-1}G\tilde{R}_w^{-1}\tilde{R}_{wx} \quad \left(\text{expansion of } \tilde{T}^{\mathrm{T}}\tilde{T} = I\right)$$

$$\hat{R}\Phi^{-1}\left(I + G\tilde{R}_w^{-1}\tilde{R}_{wx}\right) = \tilde{T}_{22}^{\mathrm{T}}\tilde{T}_{22}\hat{R}\Phi^{-1} \tag{A.3}$$

From (VI.2.29) we also obtain

$$\tilde{T}_{22}\hat{R}\Phi^{-1} = \tilde{R}$$

so that

$$\tilde{T}_{22} = \tilde{R}\Phi\hat{R}^{-1} \tag{A.4}$$

Using this in (A.3) gives

$$C_j = \Phi^{-1}\left(I + G\tilde{R}_w^{-1}\tilde{R}_{wx}\right) = \hat{R}^{-1}\hat{R}^{-\mathrm{T}}\Phi^{\mathrm{T}}\tilde{R}^{\mathrm{T}}\tilde{R} \tag{A.5}$$

or, since $R^{-1}R^{-\mathrm{T}} = P$,

$$C_j = \hat{P}_x(j)\Phi_j^{\mathrm{T}}\left[\tilde{P}_x(j+1)\right]^{-1} \tag{A.6}$$

To complete the conversion of the DMCS recursion (X.2.10) to its filter covariance form, we need $\Phi^{-1}LL^{\mathrm{T}}\Phi^{-\mathrm{T}}$ expressed in terms of filter covariances.

$$\Phi^{-1}L(\Phi^{-1}L)^{\mathrm{T}} = \left(\hat{R}^{-1}\tilde{T}_{12}^{\mathrm{T}}\right)\left(\hat{R}^{-1}\tilde{T}_{12}^{\mathrm{T}}\right)^{\mathrm{T}} \quad \text{(cf. Eq. (A.2))}$$

$$\Phi^{-1}LL^{\mathrm{T}}\Phi^{-1} = \hat{R}^{-1}\left(I - \tilde{T}_{22}^{\mathrm{T}}\tilde{T}_{22}\right)\hat{R}^{-\mathrm{T}} \quad \left(\text{because } \tilde{T}^{\mathrm{T}}\tilde{T} = I\right)$$

Using (A.4) for $\tilde{T}_{22}$ gives

$$\Phi^{-1}LL^{\mathrm{T}}\Phi^{-\mathrm{T}} = \hat{P}_x(j) - \hat{P}_x(j)\Phi_j^{\mathrm{T}}\left[\tilde{P}_x(j+1)\right]^{-1}\Phi_j\hat{P}_x(j) \tag{A.7}$$

When (A.6) and (A.7) are substituted into the DMCS recursion (X.2.10), we obtain the Rauch–Tung–Streibel smoother recursion

$$P_x^*(j) = \hat{P}_x(j) - C_j\left(\tilde{P}_x(j+1) - P_x^*(j+1)\right)C_j^{\mathrm{T}} \tag{A.8}$$

Based on the formulae developed thus far it is a simple exercise to express the smoothed state recursion (X.2.2) in a SRIF free form. From (X.2.2) we have

$$x_j^* = \Phi^{-1}(I + G\tilde{R}_w^{-1}\tilde{R}_{wx})x_{j+1}^* - \Phi^{-1}G\tilde{R}_w^{-1}\tilde{z}_w \tag{A.9}$$

and from the propagation algorithm (VI.2.29) we have

$$\tilde{z}_w = \tilde{T}_{11}z_w + \tilde{T}_{12}\hat{z}_x(j) = \tilde{T}_{11}R_w\overline{w} + \tilde{T}_{12}\hat{R}\hat{x} \tag{A.10}$$

because $z_w = R_w\overline{w}$ and $\hat{z}_x(j) = \hat{R}_x(j)\hat{x}_j = \hat{R}\hat{x}$. Again, from (VI.2.29), we obtain

$$\tilde{T}_{11}R_w - \tilde{T}_{12}\hat{R}\Phi^{-1}G = \tilde{R}_w \tag{A.11}$$

Armed with (A.10) and (A.11) we return to (A.9) and complete the derivation via the following string of equations:

$$\begin{aligned}
x_j^* &= C_j x_{j+1}^* - \Phi^{-1}G\tilde{R}_w^{-1}\tilde{z}_w && \text{(cf. Eq. (A.5))}\\
&= C_j x_{j+1}^* - \Phi^{-1}G\tilde{R}_w^{-1}(\tilde{T}_{11}R_w\overline{w} + \tilde{T}_{12}\hat{R}\hat{x}) && \text{(cf. Eq. (A.10))}\\
&= C_j x_{j+1}^* - \Phi^{-1}G\tilde{R}_w^{-1}(\tilde{R}_w\overline{w} + \tilde{T}_{12}\hat{R}\Phi^{-1}G\overline{w} + \tilde{T}_{12}\hat{R}\hat{x}) && \text{(cf. Eq. (A.11))}\\
&= C_j x_{j+1}^* - \Phi^{-1}G(\overline{w} + \tilde{R}_w^{-1}\tilde{T}_{12}\hat{R}\Phi^{-1}\tilde{x}_{j+1}) && (\text{because } \tilde{x} = \Phi\hat{x} + G\overline{w})\\
&= C_j x_{j+1}^* + \hat{x}_j - (I + \Phi^{-1}G\tilde{R}_w^{-1}\tilde{T}_{12}\hat{R})\Phi^{-1}\tilde{x}_{j+1}\\
&= \hat{x}_j + C_j x_{j+1}^* - (I - \hat{R}^{-1}\tilde{T}_{12}^{\mathrm{T}}\tilde{T}_{12}\hat{R})\Phi^{-1}\tilde{x}_{j+1} && \text{(cf. Eq. (A.2))}\\
&= \hat{x}_j + C_j x_{j+1}^* - \hat{R}^{-1}\tilde{T}_{22}^{\mathrm{T}}\tilde{T}_{22}\hat{R}\Phi^{-1}\tilde{x}_{j+1} && (\text{because } \tilde{T}^{\mathrm{T}}\tilde{T} = I)\\
&= \hat{x}_j + C_j(x_{j+1}^* - \tilde{x}_{j+1}) && \text{(cf. Eq. (A. 3))}
\end{aligned} \tag{A.12}$$

This result is (essentially) the Rauch *et al.* smoothed state recursion; its form indicates why C_j is called the smoother gain. Rauch *et al.* solved the smoothing problem for zero mean process noise ($\overline{w} = 0$), and instead of (A.12) they obtained

$$x_j^* = \hat{x}_j + C_j(x_{j+1}^* - \Phi_j\hat{x}_j)$$

This is, of course, equivalent to (A.12) since $\tilde{x}_{j+1} = \Phi_j\hat{x}_j + G\overline{w}$.

The Rauch *et al.* estimate-covariance recursions (A.12) and (A.8) are useful for theoretical purposes and have had a great impact on the smoothing technology developed since their publication. Viewed pragmatically, however, these results are uninspiring because they suggest that smoothing requires a considerable additional investment, beyond that required for filtering, both in computer storage and in computation. The recursions require that estimate and covariance histories (both predicted and filtered) be stored and that the filter predicted covariance be inverted at each step.

Estimation analysts generally distrust matrix inversions and prefer instead to use recursions to calculate the covariance inverses. This is tantamount to calculating two filters, one a covariance filter and one an information filter. Thus smoothing with the Rauch *et al.* recursions is comparable to running two filters and combining the results. An interesting observation in this regard is that one can, in fact, view the smoothing solution in terms of two such filters (cf. Fraser [3]). Fraser observed that a smoothed estimate can be obtained by combining filtered estimates; one forward (based on *a priori* statistics and all the data up to that time) and one backward (based only on the data from the terminal time back to that time).

Remark: The inference drawn from the various smoother results is that smoothing effectively requires the operation of two filters. It is, however, interesting to note (cf. Bierman [5]), that the DMCS results (2.2) and (2.10) require the least amount of computation.

One can introduce efficiencies into the Rauch *et al.* formulation and make greater use of the various quantities computed by the Kalman filter. This is the approach taken by Bierman [2] and the result, which we call the mBF smoother†, is designed to operate with Kalman gains, filter residuals, and innovation covariances in much the same way that the DMCS operates with the quantities developed by the SRIF. The efficiency of this smoothing algorithm is due in large part of the fact that estimates and covariances are not recursively computed. In their place one recursively computes adjoint variable estimates λ and adjoint covariances Λ using (A.13)–(A. 16), for $j = N - 1, \ldots, 0$.

Time Update

$$\hat{\lambda}_j = \Phi_j^{\mathrm{T}} \tilde{\lambda}_{j+1} \tag{A.13}$$

$$\hat{\Lambda}_j = \Phi_j^{\mathrm{T}} \tilde{\Lambda}_{j+1} \Phi_j \tag{A.14}$$

†The algorithm is called mBF (modified Bryson–Frazier) because the result, derived from the work of Rauch *et al.*, was found to be a more efficient form of a result reported by Bryson and Frazier (1963); see Bryson and Ho [4], Section 13.2.

Measurement Update

$$\tilde{\lambda}_j = \hat{\lambda}_j - A_j D_j^{-1}\left(\epsilon_j + D_j K_j \hat{\lambda}_j\right) \tag{A.15}$$

$$\tilde{\Lambda}_j = (I - K_j A_j)^{\mathrm{T}} \hat{\Lambda}_j (I - K_j A_j) + A_j^{\mathrm{T}} D_j^{-1} A_j \tag{A.16}$$

where the quantities in (A.15) and (A.16) are as follows:

A_j—measurement coefficients ($z_j = A_j x_j + \nu_j$),
ϵ_j—predicted filter residual, or innovation ($\epsilon_j = z_j - A_j \tilde{x}_j$),
D_j—predicted residual covariance ($D_j = A_j \tilde{P}_j A_j^{\mathrm{T}} + R_j$), and
K_j—Kalman filter gain.

At those times when estimates and/or covariances are desired they are computed from

$$x_j^* = \bar{x}_j - \bar{P}_j \bar{\lambda}_j \tag{A.17}$$

and

$$P_j^* = \bar{P}_j - \bar{P}_j \bar{\Lambda}_j \bar{P}_j \tag{A.18}$$

where the overbar represents either ~ or ^, whichever happens to be more convenient to use. Features of this mechanization include the following:

(1) Comparison of Eqs. (A.13)–(A.16) with the Kalman filter equations, Appendix II.D. and Eq. (VI.3.2), reveals a duality between the adjoint variables, λ and Λ, and the filter estimate and covariance. One can exploit this duality to rename the variables in a Kalman filter program and thus obtain a smoothing program (cf. Bierman [2]).

(2) The duality relationship between the adjoint and filter variables shows that processing measurements one component at a time is an efficient smoother mechanization (because this method is efficient for the filter algorithm).

(3) This filter adjoint–variable duality shows also that factorization algorithms can be applied. In particular, the U–D computed, sensitivity, and consider results of Chapter VII extend to the mBF adjoint recursions. We note in passing that the Potter measurement update (which involves a nontriangular square root matrix) is well suited to the adjoint problem because the propagation equation (A.14) is homogeneous. Consequently, the square root time update reduces to a single matrix multiplication.

(4) Processing data one component at a time gives rise to storage economies too. The mBF adjoint variable algorithm utilizes the filter gains K, the predicted residuals ϵ, and the predicted residual covariance D. Since data is processed one component at a time, K, ϵ, and D are an N_x vector and two scalars that are stored for each observation. Filter estimates and

covariances need only be saved for those times when estimates and/or covariances are to be exhibited.

(5) Note that covariances are not computed recursively, viz. Eq. (A.18); and this suggests that the covariance computations may be somewhat insensitive to roundoff errors, i.e., the algorithm is stable. Limited computational experience supports this hypothesis.

(6) It has been observed in Chapter II that for problems with measurements of inappreciable size the Kalman filter is quite efficient; the mBF is also efficient in this situation. In contrast, for problems involving measurement vectors of more substantial size the SRIF–SRIS and SRIF–smoothed estimate DMCS are more efficient algorithms. This is comforting because we would be hard pressed to make a general recommendation if, say, the filter algorithm were inefficient for a certain class of problems, but the companion smoother was efficient for the same class of problems.

Appendix X.B Smoother Recursions Corresponding to the Epoch State SRIF

The smoothing recursion computations degenerate almost to insignificancy when the epoch state model (VII.1.1) and the colored process noise inclusion, applied one component at a time, are used for the filtering mechanization (cf. Appendix VII.B). In this appendix we describe algorithm details that are responsible for this computational efficiency. Skillful computer mechanization of the smoothing equations considerably reduces the amount of computation and storage that is required. To better exhibit the mechanization and to avoid computer implementation misunderstandings we make heavy use of the FORTRAN method of writing over old variables. To focus attention on this one should recall (cf. Section III.3) that the expression $a := b(a)$ means "evaluate the expression $b(a)$ and store the result in the place that previously held a."

Suppose that we have smoothed computed† estimates of p_{n+1}, x_{n+1}, y, a smoothed sensitivity Sen^*_{n+1}, and a smoothed computed covariance P^{c*}_{n+1} (corresponding to p and x). These quantities are assumed to reside in

$$X^{c*} = \begin{bmatrix} p^{c*} \\ x^{c*} \end{bmatrix}, \quad y^*, \quad \text{Sen}^*, \quad P^{c*} \tag{B.1}$$

Remarks: (1) With the exception of y^* these quantities refer to time $n + 1$. When we have finished one cycle of the (backward) recursion they will refer to time n.

†Recall that "computed" estimates do not include the effects of the y bias parameters.

(2) Dimensions of the variables shown in (B.1) are $p^{c*}(N_p)$, $x^{c*}(N_x)$, $y^*(N_y)$, $\text{Sen}^*(N_{px}, N_y)$ and $P^{c*}(N_{px}, N_{px})$ (symmetric), where $N_{px} = N_p + N_x$.

The $*$ array, Eq. (VII.3.4), computed during the mapping from n to $n + 1$, is

$$\left[\; R_p^* \;\middle|\; R_{pp}^* \quad R_{px}^* \quad R_{py}^* \quad z_p^* \; \right] \equiv \left[\; \sigma^* \;\middle|\; S^* \; \right] \tag{B.2}$$

Remark: σ^* is an N_p vector and S^* is a matrix of dimensions (N_p, N_{tot}), where $N_{\text{tot}} = N_p + N_x + N_y + 1$.

This mechanization is based on Bierman's one-component-at-a-time colored process noise inclusion mechanization (cf. Appendix VII.B). To understand the significance of the columns of (B.2) we pause for a moment and review the relevant steps of Bierman's SRIF filter mechanization. At time n, after processing data, the $[{}^p_X]$ data equations are

$$\hat{S}_p p_n + \hat{S}_x x_n + \hat{S}_y y = \hat{S}_z - \hat{S}_\nu, \tag{B.3}$$

where $\hat{S}_\nu \in N(0, I)$. The mapping (VII.B.3) from time n to time $n + 1$ is arranged as

$$\left.\begin{aligned} \chi_{k+1} &= \chi_k + V_p(k) p_n(k) \\ p_{n+1}(k) &= m_k p_n(k) + w_n(k) \end{aligned}\right\} k = 1, \ldots, N_p \tag{B.4}$$

where $\chi_1 = x_n$, and $\chi_{N_p+1} = x_{n+1}$. The propagation uses the mapping (B.4), carried off in N_p steps, together with (B.3), which represents the *a priori* estimate-covariance information. Since only one component of p changes at each step, the state vector for this process is of the form

$$X_{k+1} = \begin{bmatrix} p_{n+1}(1) \\ \vdots \\ p_{n+1}(k) \\ p_n(k+1) \\ \vdots \\ p_n(N_p) \\ \hline \chi_{k+1} \end{bmatrix} = \begin{bmatrix} X_{k+1}(1) \\ \cdot \\ \cdot \\ \cdot \\ X_{k+1}(N_p) \\ \hline X_{k+1}(N_p + 1) \end{bmatrix} \tag{B.5}$$

Remark: Observe that

$$X_1 = \begin{bmatrix} p_n \\ x_n \end{bmatrix} \quad \text{and} \quad X_{N_p} = \begin{bmatrix} p_{n+1} \\ x_{n+1} \end{bmatrix}$$

Note also that the $(N_p + 1)$st component of X_{k+1}, $X_{k+1}(N_p + 1)$, is itself a vector and has dimension N_x.

The SRIF computer mechanization (cf. Appendix VII.B) is arranged so that the array (B.2) represents

$$\sigma^*(k)p_n(k) + \sum_{j=1}^{N_p-k} S^*(k,j)p_n(k+j) + \sum_{j=N_p-k+1}^{N_p} S^*(k,j)p_{n+1}(j - N_p + k)$$
$$+ \sum_{j=N_p+1}^{N_{px}} S^*(k,j)x_{k+1}(j - N_p) + \sum_{j=1}^{N_y} R^*_{py}(k,y)y(j) = z^*_p(k) - \nu^*_p(k) \tag{B.6}$$

for $k = 1, \ldots, N_p$. To express (B.6) in terms of X_{k+1} we redefine the kth row of S^*,

$$[S^*(k,1) \cdots S^*(k,N_p)]$$
$$:= [\underbrace{S^*(k,1) \cdots S^*(k,N_p-k)}_{N_p - k \text{ terms}}\overbrace{S^*(k,N_p-k+1) \cdots S^*(k,N_p)}^{k \text{ terms}}] \tag{B.7}$$

i.e., the last k elements become the first k elements. After scaling the entire kth row of this redefined S^*, Eq. (B.9), we rewrite (B.6) in terms of X_{k+1}:

$$\lambda := 1/\sigma^*(k) \tag{B.8}$$

$$S^*(k,j) := S^*(k,j)\lambda, \qquad j = 1, \ldots, N_{\text{tot}} \tag{B.9}$$

$$p_n(k) + \sum_{j=1}^{N_{px}} S^*(k,j)X_{k+1}(j) + \sum_{j=1}^{N_y} R^*_{py}(k,j)y(j) = z^*_p(k) - \lambda\nu^*_p(k) \tag{B.10}$$

for $k = 1, \ldots, N_p$.†

†Note that the scaling, (B.9), results in redefined R^*_{py} and z^*_p elements.

The smoothing procedure involves, so to speak, undoing the mapping procedure (B.10); therefore (B.10) is utilized in reverse order. The backward recursion to be used for smoothing involves (B.10) together with the χ mapping of Eq. (B.4):

$$\left.\begin{aligned} p_n(k) = X_k(k) &= -\sum_{j=1}^{N_{px}} S^*(k,j)X_{k+1}(j) \\ &\quad -\sum_{j=1}^{N_y} R^*_{py}(k,j)y(j) + z^*_p(k) - \lambda v^*_p(k) \\ X_k(i) &= X_{k+1}(i), \qquad i = 1, \ldots, N_p, \quad i \neq k \\ X_k(N_p+1) &= X_{k+1}(N_p+1) - V_p(k)X_k(k) \end{aligned}\right\} \tag{B.11}$$

for $k = N_p, N_p - 1, \ldots, 1$.

Remark: Recall that $V_p(k)$ is the kth column of the V_p matrix at time n (cf. Eq. (VII.B.3)).

Equations (B.11) are the relationships on which our smoother mechanization is based. To separate out the effects of the y bias parameters we rearrange the equations into two separate recursions (cf. Eqs. (3.4)–(3.7)):

$$\left.\begin{aligned} X_k^{c*}(k) &= z^*_p(k) - \sum_{j=1}^{N_{px}} S^*(k,j)X_{k+1}^{c*}(j) - \lambda v^*_p(k) \\ X_k^{c*}(i) &= X_{k+1}^{c*}(i), \qquad i = 1, \ldots, N_p, \quad i \neq k \\ X_k^{c*}(N_p+1) &= X_{k+1}^{c*}(N_p+1) - V_p(k)X_k^{c*}(k) \end{aligned}\right\} \tag{B.12}$$

and

$$\left.\begin{aligned} X_k^{y*}(k) &= -\sum_{j=1}^{N_{px}} S^*(k,j)X_{k+1}^{y*}(j) - \sum_{j=1}^{N_y} R^*_{py}(k,j)y(j) \\ X_k^{y*}(i) &= X_{k+1}^{y*}(i), \qquad i = 1, \ldots, N_p, \quad i \neq k \\ X_k^{y*}(N_p+1) &= X_{k+1}^{y*}(N_p+1) - V_p(k)X_k^{y*}(k) \end{aligned}\right\} \tag{B.13}$$

X^c involves the effects of the noise and the *a priori* statistics, and X^y reflects only the effects due to the presence of the y bias parameters; X^c and X^y are independent and

$$X = X^c + X^y \tag{B.14}$$

In the algorithms to follow we initiate our recursions at time $n + 1$ and step back N_p steps to time n.

For $k = N_p, N_p - 1, \ldots, 1$ perform steps (a)–(d).

(a) Evaluate Eqs. (B.7)–(B.9) so that the kth row of S^* has been rearranged and scaled.

(b) Update the computed estimate X^{c^*}:

$$\left.\begin{aligned} \delta &:= \sum_{j=1}^{N_{px}} S^*(k, j) X^{c^*}(j) \\ X^{c^*}(k) &:= S^*(k, N_{\text{tot}}) - \delta \\ X^{c^*}(N_p + 1) &:= X^{c^*}(N_p + 1) - V_p(k) X^{c^*}(k) \end{aligned}\right\} \tag{B.15}$$

Remark: Each time Eq. (B.15) is applied, another component of $p_n^{c^*}$ is updated. It is only after all N_p applications of these equations that $X^{c^*}(N_p + 1) = X$ represents the computed estimate at time n.

(c) Update the sensitivity Sen*:

$$\left.\begin{aligned} \delta &:= \sum_{j=1}^{N_{px}} S^*(k, j) \operatorname{Sen}^*(j, m) \\ \operatorname{Sen}^*(k, m) &:= -S^*(k, N_{px} + m) - \delta \\ \operatorname{Sen}^*(N_p + 1, m) &:= \operatorname{Sen}^*(N_p + 1, m) \\ &\quad - V_p(k) \operatorname{Sen}^*(k, m) \end{aligned}\right\} \quad m = 1, \ldots, N_y \tag{B.16}$$

Remarks: (1) The Sen* equations are virtually the same as the X^{c^*} equations (B.15) because the columns of Sen* are scaled estimates of X.

(2) Our thoughts are on computer implementation, and this motivates us to write $S^*(k, N_{\text{tot}})$ and $S^*(k, N_{px} + m)$ instead of $z_p^*(k)$ and $R_{py}(k, m)$.

(d) Update the computed covariance matrix P^{c^*}:

It is easy to see that Eqs. (B.12) give rise to the covariance recursion

$$P^{c^*} := \Psi P^{c^*} \Psi^{\mathrm{T}} + \lambda^2 L L^{\mathrm{T}} \tag{B.17}$$

where

$$\Psi := \begin{array}{c} \begin{array}{cccc} \overbrace{\qquad}^{k-1} & \overbrace{\qquad}^{1} & \overbrace{\qquad}^{N_p - k} & \overbrace{\qquad}^{N_x} \end{array} \\ \left[\begin{array}{c|c|c|c} I & 0 & 0 & 0 \\ \hline h_1 & h_2 & h_3 & h_4 \\ \hline 0 & 0 & I & 0 \\ \hline -Vh_1 & -Vh_2 & -Vh_3 & I - Vh_4 \end{array}\right] \begin{array}{l} \}k-1 \\ \}1 \\ \}N_p - k \\ \}N_x \end{array} \end{array} \tag{B.18a}$$

$$L := \begin{bmatrix} 0 \\ 1 \\ 0 \\ -V \end{bmatrix} \begin{matrix} \}k-1 \\ \}1 \\ \}N_p - k, \\ \}N_x \end{matrix} \qquad V := V_p(k) \tag{B.18b}$$

$$h = [h_1 \quad h_2 \quad h_3 \quad h_4] := -[S^*(k, 1) \cdots S^*(k, N_{px})] \tag{B.18c}$$

The special form of Ψ (it depends on only two vectors) suggests that partitioning can be exploited to significantly reduce the computation of $\Psi P^{c^*} \Psi^T$. Equations for the nonredundant upper triangular elements of the updated P^{c^*} are as follows [(B.19)–(B.24)]:

$$z := P^{c^*} h^T$$

This equation is implemented in terms of the upper triangular elements of P^{c^*} and the elements of $S^*(k, \cdot)$, i.e.,

$$z(i) := -\sum_{m=1}^{i} P^{c^*}(m, i) S^*(k, m) - \sum_{m=i+1}^{N_{px}} P^{c^*}(i, m) S^*(k, m), \qquad i = 1, \ldots, N_{px} \tag{B.19}$$

The covariance updating is determined, in the main, from z.

$$\left.\begin{aligned} P^{c^*}(m, k) &:= z(m) \\ P^{c^*}(m, N_{pj}) &:= P^{c^*}(m, N_{pj}) - z(m) V_p(j, k), \\ & \qquad j = 1, \ldots, N_x \end{aligned}\right\} m = 1, \ldots, k-1 \tag{B.20}$$

$$P^{c^*}(k, k) := \lambda^2 - \sum_{j=1}^{N_{px}} S^*(k, j) z(j) \tag{B.21}$$

$$P^{c^*}(k, j) := z(j), \qquad j = k+1, \ldots, N_p \tag{B.22}$$

$$\left.\begin{aligned} P^{c^*}(k, N_p + m) &:= z(N_p + m) - P^{c^*}(k, k) V_p(m, k) \\ P^{c^*}(j, N_p + m) &:= P^{c^*}(j, N_p + m) - z(j) V_p(m, k), \\ & \qquad j = k+1, \ldots, N_p \end{aligned}\right\} m = 1, \ldots, N_x \tag{B.23}$$

Remark: Note that the result of (B.21) is used in (B.23).

$$P^{c^*}(N_{pj}, N_{pm}) := P^{c^*}(N_{pj}, N_{pm}) - V_p(j, k) P^{c^*}(k, N_{pm}) - z(N_{pj}) V_p(m, k)$$
$$N_{pj} = N_p + j, \qquad N_{pm} = N_p + m, \qquad j = 1, \ldots, m, \qquad m = 1, \ldots, N_x \tag{B.24}$$

Remark: Note that the result of (B.23) is used in (B.24).

CAUTION

Rearranging the order of these equations can change the results!

The reader may be inclined to view Eqs. (B.19)–(B.24) with trepidation and might prefer, instead, to evaluate the more compact covariance equation (B.17). Our reasons for replacing (B.17) with (B.19)–(B.24) include:

(1) The use of (B.17) directly would require the formation of Ψ and a temporary storage area for ΨP^{c^*}. We are able to avoid additional matrix storage (except for the work vector z). Thus storage is reduced.

(2) Our mechanization involves but a fraction of the computations required to evaluate (B.17).

(3) These two reasons are not, in themselves, convincing arguments as typical orbit determination navigation problems involve less than 15 stochastic variables (N_p) and 6 state variables (N_x). For matrices with such paltry dimensions storage and computation are not critical issues. It turns out, however, that the amount of computer coding needed to implement (B.19)–(B.24) is about the same as that needed to implement the seemingly more compact (B.17). When it is realized that both implementations require the same effort to implement our method is preferable.

(4) Although our mechanization appears somewhat cryptic, its derivation involves only the application of elementary matrix partitioning and some skillful grouping of terms.

The recursions just described carry X^{c^*}, Sen*, and P^{c^*} from step $n + 1$ back to step n; and the recursions are repeated for $n = N - 1, N - 2, \ldots, 0$. The basic quantities can be used to compute variable-order smoothed estimates and corresponding estimate error covariances (based on different sets of y parameters). In addition one can compute smoothed consider covariances for smoothed estimates that do not include the effects of the y (consider) parameters. An important feature of our development is that the formulae for these calculations are precisely the same as those used for the filtering solution (cf. Chapters VII and IX).

References

[1] Rauch, H. E., Tung, F., and Striebel, C. T., Maximum likelihood estimates of linear dynamic systems, *AIAA J.* **3**, No. 8, 1445–1450 (August 1965).

This paper is one of the classics in the field of linear smoothing. It is frequently used as a starting point for the development of new algorithms and analyses.

[2] Bierman, G. J., Fixed interval smoothing with discrete measurements, *Int. J. Control* **18**, No. 1, 65–75 (1973).

The main point of this paper is a computationally efficient smoothing algorithm that is predicated on the Kalman filter.

[3] Fraser, D. C., A new technique for the optimal smoothing of data, MIT Instrumentation Lab., Rep. No. T-474, Cambridge, Massachusetts (January 1967).

This is Fraser's Ph.D thesis. It contains an abundance of useful relations for filtering and smoothing.

[4] Bryson, A. E., and Ho, Y. C., "Applied Optimal Control." Ginn (Blaisdell), Waltham, Massachusetts (1969).

In this text estimation is treated as an optimization problem. The authors draw upon their extensive optimization experience to derive smoothing algorithms as the solution to a two point boundary value optimization problem in the continuous case and as a dynamic programming problem in the discrete case. This approach is a refreshing change from the more commonly used Baysian technique.

[5] Bierman, G. J., Sequential square root filtering and smoothing of discrete linear systems, *Automatica* **10**, 147–158 (March 1974).

Much of the material from Section 2 is described in this expository paper. It contains an appendix which compares various smoothing algorithms in terms of operation counts. Since arithmetic operation counts are dependent upon specific computer mechanizations, the paper includes detailed algorithm formulations.

Bibliography

Deciding which references to include in a bibliography is difficult because of a tendency toward copiousness. We suppressed this impulse and decided that no attempt would be made toward compiling a complete, or even an extensive, list of references on estimation theory. The excellent works by Deutsch (1965), Sorenson (1970), Sage and Melsa (1971), Mendel and Geiseking (1971), and Kailath (1974) provide an ample supply of references on the various aspects of estimation. We have chosen to include only those works referenced in the text or which we believe supplement topics that have been discussed there. We apologize to those of our colleagues whose relevant publications are not referenced.

Books, Theses, and Reports

Albert, A. (1972), "Regression and the Moore–Penrose Pseudoinverse." Academic Press, New York.

Aoki, M. (1967), "Optimization of Stochastic Systems—Topics in Discrete Time Systems." Academic Press, New York.

Battin, R. H. (1964), "Astronautical Guidance." McGraw-Hill, New York.

Bryson, A. E., and Ho, Y. C. (1969), "Applied Optimal Control." Ginn (Blaisdell), Waltham, Massachusetts.

Boullion, T. L., and Odell, P. L., (1966), An introduction to the theory of generalized matrix invertibility, NASA Rep. No. NASA CR-62058.

Bucy, R. S., and Joseph, P. D. (1968), "Filtering for Stochastic Processes with Applications to Guidance." Wiley, New York.

Curkendall, D. W. (1972), Problems in estimation theory with applications to orbit determination, UCLA School Eng. Appl. Sci. Rep. No. UCLA-ENG-7225.

Deutsch, R. (1965), "Estimation Theory." Prentice-Hall, Englewood Cliffs, New Jersey.

Fadeev, D.K., and Fadeeva, V.N. (1963), "Computational Methods of Linear Algebra," Freeman, San Francisco.

Fraser, D. C. (1967), A new technique for the optimal smoothing of data, MIT Ph.D. Thesis, also appeared as MIT Instrumentation Lab. Rep. No. T-474, Cambridge, Massachusetts.

Householder, A. S. (1964), "The Theory of Matrices in Numerical Analysis." Ginn (Blaisdell), Waltham, Massachusetts.

Jazwinski, A. H. (1970), "Stochastic Processes and Filtering Theory." Academic Press, New York.

Kalman, R. E. (1963), New methods in Wiener filtering theory, *In Proc. 1st Symp. Eng. Appl. Random Function Theory Probability* (J. L. Bogdanoff and F. Kozin, eds.), pp. 270–388. Wiley, New York.

Kaminski, P. G. (1971), Square root filtering and smoothing for discrete processes, Ph.D. Dissertation, Dept. Aeronaut. Astronaut., Stanford Univ., Stanford, California.

Lawson, C. L., and Hanson, R. J. (1974), "Solving Least Squares Problems." Prentice-Hall, Englewood Cliffs, New Jersey.

Leondes, C. T. (ed.) (1970), Theory and applications of Kalman filtering, NATO Advisory Group for Aerospace Res. Develop., AGARDograph 139. (Available from the Clearinghouse for Federal Scientific and Technical Information, Springfield, Virginia, 22151, Rep. No. AD 704306.)

Meditch, J. S. (1969), "Stochastic Optimal Linear Estimation and Control." McGraw-Hill, New York.

Morf, M. (1974), Fast algorithms for multivariable systems, Ph.D. Dissertation, Dept. Elect. Eng., Stanford Univ., Stanford, California.

Noble, B., (1969), "Applied Linear Algebra." Prentice-Hall, Englewood Cliffs, New Jersey.

Sage, A. P. and Melsa, J. L., (1971), "Estimation Theory with Applications to Communications and Control." McGraw-Hill, New York.

Schmidt, S. F., (1970), Computational techniques in Kalman filtering, *In* "Theory and Applications of Kalman Filtering" (C.T. Leondes, ed.), Advisory Group for Aerospace Res. Develop., AGARDograph 139, 67–103.

Sorenson, H. W., (1965), Kalman filtering techniques, *Adv. Control Systems* **3**, 219–292.

Thornton, C. L., (1976), Triangular covariance factorizations for Kalman filtering, Ph.D. Dissertation, UCLA School of Engineering, Dept. of Systems Science.

Turnbull, H. W., and Aitken, A. C., (1930), "An Introduction to the Theory of Canonical Matrices." Blackie, London and Glasgow.

Wedderburn, H. H. M. (1934), "Lectures on Matrices." Amer. Math. Soc., Providence, Rhode Island.

Wilkinson, J. (1965a), Error analysis of transformations based on the use of matrices of the form $I - 2ww^T$, *In* "Errors in Digital Computation" (L. B. Rall, ed.), Volume II, pp. 77–101. Wiley, New York.

Wilkinson, J. (1965b), "The Algebraic Eigenvalue Problem." Clarendon Press, Oxford.

Papers

Agee, W. S., and Turner, R. H. (1972), Triangular decomposition of a positive definite matrix plus a symmetric dyad with application to Kalman filtering, White Sands Missile Range Tech. Rep. No. 38.

Andrews, A. (1968), A square root formulation of the Kalman covariance equations, *AIAA J.* **6**, 1165–1166.

Athans, M. (1968), The matrix minimum principle, *Inform. Control*, **11**, 592–606.

Athans, M., and Schweppe, F. C. (1965), Gradient matrices and matrix calculations, MIT Lincoln Lab. Tech. Note 1965-53, Lexington, Massachusetts.

Athans, M., Wishner, R. P., and Bertolini, A. (1968), Suboptimal state estimation for continuous-time nonlinear systems from discrete noisy measurements, *IEEE Trans. Automatic Control*, **AC-13**, No. 5, 504–514.

Bellantoni, J. F. and Dodge, K. W. (1967), Square root formulation of the Kalman–Schmidt filter, *AIAA J.* **5**, 1309–1314.

Bennett, J. M. (1965), Triangular factors of modified matrices, *Num. Math.* **7**, 217–221.

Bierman, G. J. (1973a), Fixed interval smoothing with discrete measurements, *Int. J. Control* **18**, No. 1, 65–72.

Bierman, G. J. (1973b), A comparison of discrete linear filtering algorithms, *IEEE Trans. Aero. Elect. Systems* **AES-9**, No. 1, 28–37.

Bierman, G. J. (1974a), Sequential square root filtering and smoothing of discrete linear systems, *Automatica* **10**, 147–158.

Bierman, G. J. (1974b), On the application of discrete square-root information filtering, *Int. J. Control* **20** (3), 465–477.

Bierman, G. J. (1975a), Measurement updating using the *U–D* factorization, *Proc. IEEE Conference on Decision and Control, Houston, Texas*, pp. 337–346.

Bierman, G. J. (1975b), The treatment of bias in the square-root information filter/smoother, *J. Optim. Theory Appl.* **16**, (1-2), 165–178.

Bierman, G. J., and Rourke, K.H. (1972), Proving a folk theorem about state variable augmentation to evaluate dynamic mismatch, *JPL Tech. Mem.* No. TM391-336.

Björck, A. (1967), Solving least squares problems by Gram–Schmidt orthogonalization, *BIT* **7**, 1–21.

Bryson, A. E., and Frazier, M. (1963), Smoothing for linear and nonlinear dynamic systems, *Proc. Optimum Systems Synthesis Conf.* U. S. Air Force Tech. Rep. ASD-TDR-063-119.

Businger, P., and Golub, G. H. (1965), Linear least squares solutions by Householder transformations, *Num. Math.* **7**, 269–276.

Carlson, N. A. (1973) Fast triangular factorization of the square root filter, *AIAA J.* **11**, No. 9, 1259–1265.

Cox, H. C. (1964), Estimation of state variables and parameters for noisy dynamic systems, *IEEE Trans. Automatic Control*, **AC-9**, No. 1, 5–12.

Christensen, C. S., and Reinbold, S. J (1974), Navigation of the Mariner 10 spacecraft to Venus and Mercury, *AIAA Mech. Control Flight Conf., Anaheim, California* AIAA paper No. 74-844.

Curkendall, D. W. (1970), Toward the design of more rational filters— a generalized consider option, JPL Tech Memo. No. TM 391–70.

Dyer, P. and McReynolds, S. (1969), Extension of square-root filtering to include process noise, *J. Opt. Theory and Appl.* **3**, No. 6, 444–459.

Friedland, B. (1969), Treatment of bias in recursive filtering, *IEEE Trans. Automatic Control* **AC-14**, No. 4, 359–367.

Gentleman, W. M. (1973), Least squares computations by Givens transformations without square roots, *J. Inst. Math. Appl.* **12**, 329–336.

Gill P. E., Golub, G. H., Murray, W., and Saunders, M. A. (1974), Methods for modifying matrix factorizations, *Math. Comp.* **28**, 505–535.

Golub, G. H. (1965), Numerical methods for solving linear least-squares problems, *Num. Math.* **7**, 206–216.

Griffin, R. E., and Sage, A. P. (1969), Sensitivity analysis of discrete filtering and smoothing algorithms, *AIAA J.* **7**, No. 10, 1890–1897.

Hanson, R. J., and Lawson, C. L. (1969), Extensions and applications of the Householder algorithm for solving linear least squares problems, *Math. Comp.* **23**, No. 108, 787–812.

Ho, Y. C. (1963), On the stochastic approximation method and optimal filtering theory, *J. Math. Anal. Appl.* **6**, 152–154.

Householder, A. H. (1958), Unitary triangularizations of a nonsymmetric matrix, *J. Assoc. Comp. Mach.* **5**, 339–342.

Jazwinski, A. H. (1973), Nonlinear and adaptive estimation in re-entry, *AIAA J.* **11**, No. 7, 922–926.

Jordan, T.L. (1968), Experiments on error growth associated with some linear least-squares procedures, *Math. Comp.* **22**, 579–588.

Kailath, T. (1974), A view of three decades of linear filtering theory, *IEEE Trans. Information Theory* **IT-20** No. 2, 146–181.

Kaminski, P. G., and Bryson, A. E. (1972), Discrete square root smoothing, *Proc. 1972 AIAA Guidance and Control Conf.*, AIAA Paper No. 72-877.

Kaminski, P. G., Bryson, A. E., and Schmidt, S. F. (1971), Discrete square root filtering: a survey of current techniques, *IEEE Trans. Automatic Control* **AC-16**, No. 6, 727–735.

McReynolds, S. R. (1969), The sensitivity matrix method for orbit determination error analysis, with applications to a Mars orbiter, NASA Space Programs Summary, Vol. II, pp. 37–56.

Mendel, J. M., and Gieseking, D. L. (1971), Bibliography on the linear quadratic Gaussian problem, *IEEE Trans. Automatic Control* **AC-16**, No. 6, 847–869.

Morf, M., and Kailath, T. (1974), Square-root algorithms for least squares estimation, *IEEE Trans. Automatic Control* **AC-20**, No. 4, 483–497.

Nishimura, T. (1966), On the *a priori* information in sequential estimation problems, *IEEE Trans. Automatic Control* **AC-11**, No. 2, 197–204.

Nishimura, T., and Nead, M. (1973), Athena filter sequential orbit determination program with general evaluation capability, JPL Tech. Rep. No. 900–605.

Rauch, H. E., Tung. F., and Striebel, C. T. (1965), Maximum likelihood estimates of linear dynamic systems, *AIAA J.* **3**, No. 8, 1445–1450.

Schlee, F. H., Standish, C. J., and Toda, N. F. (1967), Divergence in the Kalman filter, *AIAA J.* **5**, 1114–1120.

Schmidt, S. F. (1972), Precision navigation for approach and landing operations, *Proc. Joint Automatic Control Conf., Standford Univ.* 455–463.

Schmidt, S. F., Weinberg, J. P., and Lukesh, J. S. (1968), Case study of Kalman filtering in the C-5 aircraft navigation system, *Joint Automatic Control Conf., Univ. of Michigan* 59–109.

Smith, G. L., Schmidt, S. F., and McGee, L. A. (1962), Application of statistical filtering to the optimal estimation of position and velocity on-board a circumlunar vehicle, NASA Ames Research Center Rep. No. NASA TND-1205.

Soong, T. T. (1965), On *a priori* statistics in minimum variance estimation problems, *J. Basic Eng.* 87D, 109–112.

Sorenson, H. W. (1967), On the error behavior in linear minimum variance estimation problems, *IEEE Trans. Automatic Control*, **AC-12**, No. 5, 557–562.

Sorenson, H. W. (1970), Least squares estimation: from Gauss to Kalman, *IEEE Spectrum* **7**, 63–68.

Thornton, C. L., and Bierman, G. J. (1975), Gram–Schmidt algorithms for covariance propagation, *Proc. IEEE Conference on Decision and Control, Houston, Texas*, pp. 489–498.

Toda, N. F., Schlee, F. H., and Obsharsky, P. (1969), Region of Kalman filter convergence for several autonomous navigation modes, *AIAA J.* **7**. No. 4, 622–627.

Wampler, R. H. (1970), A report on the accuracy of some widely used least squares procedures, *J. Amer. Stat. Assoc.* **65**, (330), pp. 549–565.

Widnall, W. S. (1973), Optimal filtering and smoothing simulation results for Cirus inertial and precision ranging data, *1973 AIAA Guidance and Control Conf.*, AIAA Paper No. 73–872.

Willman, W.W. (1969), On the linear smoothing problem, *IEEE Trans. Automatic Control* **AC-14**, 116–117.

Wilkinson, J. H. (1961), Error analysis of direct methods of matrix inversion, *J. Assoc. Comput. Mach.* **8**, 281–330.

Woodbury, M. (1950), Inverting modified matrices, Memorandem Rep. 42, Statistical Research Group, Princeton, New Jersey.

Index

I

K

L

M

N

O

P

V

W

A
B 7
C 8
D 9
E 0
F 1
G 2
H 3
I 4
J 5